Vibration Control of Structures

Edited by Cyril Fischer and Jiří Náprstek

Published in London, United Kingdom

IntechOpen

Supporting open minds since 2005

Vibration Control of Structures
http://dx.doi.org/10.5772/intechopen.91569
Edited by Cyril Fischer and Jiří Náprstek

Contributors
Yucheng Liu, Ge He, Leyla Fali, Khaled Zizouni, Ismail Khalil Bousserhane, Mohamed Djermane, Abdelkrim Saidi, Abella El Kabouss, El Hassan Zerrik, Kenneth C. Crawford, Cyril Fischer, Jiří Náprstek

Notice
Statements and opinions expressed in the chapters are these of the individual contributors and not necessarily those of the editors or publisher. No responsibility is accepted for the accuracy of information contained in the published chapters. The publisher assumes no responsibility for any damage or injury to persons or property arising out of the use of any materials, instructions, methods or ideas contained in the book.

First published in London, United Kingdom, 2023 by IntechOpen
IntechOpen is the global imprint of INTECHOPEN LIMITED, registered in England and Wales, registration number: 11086078, 5 Princes Gate Court, London, SW7 2QJ, United Kingdom
Printed in Croatia

British Library Cataloguing-in-Publication Data
A catalogue record for this book is available from the British Library

Additional hard and PDF copies can be obtained from orders@intechopen.com

Vibration Control of Structures
Edited by Cyril Fischer and Jiří Náprstek
p. cm.
Print ISBN 978-1-83968-888-1
Online ISBN 978-1-83968-889-8
eBook (PDF) ISBN 978-1-83968-890-4

We are IntechOpen,
the world's leading publisher of
Open Access books
Built by scientists, for scientists

6,200+
Open access books available

168,000+
International authors and editors

185M+
Downloads

156
Countries delivered to

Our authors are among the

Top 1%
most cited scientists

12.2%
Contributors from top 500 universities

Interested in publishing with us?
Contact book.department@intechopen.com

Numbers displayed above are based on latest data collected.
For more information visit www.intechopen.com

Meet the editors

Cyril Fischer is a senior researcher at the Institute of Theoretical and Applied Mechanics of the Czech Academy of Sciences. He graduated from Charles University in Prague and in 2002 he received his Ph.D. in numerical mathematics from the Faculty of Mathematics and Physics of the same university. His thesis "Numerical Methods in Stochastic Mechanics" was awarded the Babuška Prize for the best work in scientific computing. His research interests include computational mechanics, stability problems, dynamic systems, and random processes.

Jiří Náprstek obtained his DSc degree in engineering mechanics from the Faculty of Civil Engineering, Czech Technical University in 1966, and was awarded his Ph.D. in 1972. During his career, he has studied various aspects of stochastic and nonlinear dynamics, dynamic stability, flow-induced vibrations, effects of inertial moving load, and other non-selfadjoint systems. Dr. Náprstek is the chairman of the Czech Society of Mechanics. He has published about 400 papers in scientific journals and proceedings of prestigious conferences, and has co-authored four monographs and coordinated five edited books. In recognition of his scientific activities, he has been awarded a number of Czech and international awards.

Contents

Preface

This book provides the reader with a comprehensive overview of the state of the art in vibration control and safety of structures, in the form of an easy-to-follow, article-based presentation that focuses on selected major developments in this critically important area.

Safety, reliability and long life of structures are the key points in engineering industries. Although many criteria for structural reliability exist, resistance to various types of dynamic loading is perhaps decisive among them. Thanks to progress in basic research in theoretical and experimental disciplines as well as in industrial development, protection against external dynamic influences need not only be passive, but can be applied through many active control systems. Consequently, the vibration control of structures is a crucial aspect of protection against sudden dynamic forces. Although the widespread introduction of digital control has meant phasing out of passive vibration diminution devices, both remain suitable and have a part to play in specific situations.

Structural vibration control is designed to suppress and control any unfavorable vibration due to dynamic forces that could alter the performance of the structure. Although many vibration control schemes have been investigated so far, questions involving their practical application, such as the use of advanced optimization techniques to control the vibration of structures, require further study. At the same time, it is necessary to realize that external environmental excitation processes are not only of a deterministic type, but are characterized by strong random processes. The two types of phenomena usually combine to form a complicated dynamic system, especially when the structure has to be considered in a nonlinear state. Such assignments require an analysis of the dynamic stability of the structure interacting with excitation sources.

In order to cover as much of the discipline as possible, the chapters of this book range from an enumeration of typical dynamic processes encountered in engineering practice to various styles of control in particular cases, highlighting the specific response processes of individual dynamic systems.

The field of vibration control of structures is, of course, much broader than the scope of presentation allowed for in this book. However, its contents represent a selection of typical topics discussed in this domain at the level of the basis of rational dynamics itself and applications in engineering practice, typified by the interaction of civil and mechanical engineering, with a possible overlap into theoretical and experimental physics.

In order to be successful in control and general management of the dynamic effects endangering civil engineering structures, it is necessary to evaluate statistics of the most serious events either of natural or operating origin. The first chapter, therefore, presents a balanced overview of structures and the causes of their failure due to disasters caused by the dynamic effects of wind, traffic, or earthquakes, due to insufficient knowledge of the dynamic behavior of structures, their complicated long-term interaction with environmental processes or material reliability, and, last but not least, due to inadequacy of standards and codes. As a consequence of these incidents, there have been significant changes in the codes and design philosophy of bridge construction in recent decades. The chapter illustrates progress in bridge engineering due to scientific and technological advances concerning the influence of wind, seismicity and heavy traffic. As the author points out, further research is still needed to mitigate the long-term effects of vibration and material degradation on the performance and integrity of bridge structures.

Chapter 2 studies a fractional distributed optimal (or sub-optimal) control for a class of infinite-dimensional parabolic bilinear systems evolving in a spatial domain Ω by distributed controls depending on the control operator. The main efficiency of the operator follows from a fractional spatial derivative of the Riemann–Liouville type. Using Fréchet differentiability, the existence of an optimal control depending on both time and space is emphasized. In principle, a quadratic function is minimized, which accounts for the deviation between the desired and the achieved state. Then, the characterizations of optimally distributed control for different admissible control sets are given. The chapter shows the importance of a strong theoretical background in dynamic models, particularly in the optimizing process, provided it is used in practice for reliable control. The authors developed and tested an algorithm materializing the above theoretical derivation. Subsequent simulations illustrate that the previous theoretical results are meaningful and can provide stable and practically applicable results.

A very interesting device that can serve to reduce structural vibrations due to various external shocks is based on the dry sliding phenomenon. Generally in physics and engineering, sliding represents both positive and negative effects with respect to the reliability and dynamic character of the system. For instance, bowed musical instruments are based on a complicated sliding force at the bow–string contact, which decreases with the velocity of the bow. In engineering, on the other hand, it can be understood as a very effective principle of energy absorption over a very large range of frequencies and amplitudes of relevant oscillations. This makes it an excellent candidate for various damping devices applicable in environments where excitations are extremely unpredictable in the frequency spectrum and amplitude content, e.g., anti-seismic facilities. These factors led to the invention in the mid-20th century of the sliding mode controller as an effective nonlinear controller for structures in seismic engineering, piping vibration damping, etc. However, practical implementation revealed that the sliding-based controller suffers from low sensitivity to uncertainties and other system variations due to chattering effects. Chattering is a harmful phenomenon because it leads to low control accuracy, high wear of moving mechanical parts, and high thermal losses in power circuits. In the first phase of the development of sliding mode control theory, the chattering was the main obstacle to its implementation. Thanks to the law of adaptation, this shortcoming was overcome by dynamically adapting the controller parameters depending on the system changes. Chapter 3 describes relevant research, numerical simulations and experimental measurements.

Another area that has seen intensive testing of dynamic effects is the suspension systems of softly sprung vehicles. In Chapter 4, a low-cost, customized, and effective damper dynamometer is constructed using computer-aided design and finite element analysis to measure the properties of suspension dampers used in a racing car. The chapter presents an excellent example of the advanced engineering process in the field, from construction design to race-track testing. Particularly inspiring is the description of the development of special equipment that had to meet strict requirements given the conditions in which it operates. The authors carefully follow the entire path of balanced design, manufacturing, testing and interpretation. Chapter 5 provides a typical example of complex nonlinear dynamic processes occurring in a non-conventional spherical absorber. Although passive absorbers have been investigated theoretically and experimentally many times and are commonly installed in practice, there are still many gaps in the information about new and progressive types. It is important to note that inappropriate configuration of a vibration absorber can not only reduce its efficiency but can even result in its negative influence. This is quite often the case, for example, with a passive or semi-active pendulum absorber when its spatial character is neglected. The system is strongly nonlinear and the interaction of the horizontal response components gives rise to complicated effects that can lead to various forms of stability loss.

A more sophisticated system, presented in the final chapter, is a ball-shaped absorber moving in a spherical cavity. This arrangement allows for many modifications. For instance, the cavity may be elliptical in shape, allowing the absorber to possess different eigen-frequencies in the principal directions, which is the usual disposition when the vibrations of a building require reduction. Another variant is a cylindrical absorber, as required for damping the horizontal vibration of a bridge deck. The spherical ball absorber is even more sensitive to dynamic nonlinear effects that cannot be avoided in the theoretical modeling of this system. Particularly interesting may also be the demonstration of Gibbs' principle of constructing the governing dynamic differential system, which in certain configurations leads to a more efficient system than that resulting from the direct application of the Hamiltonian principle to obtain a Lagrangian system.

Jiří Náprstek and Cyril Fischer
Institute of Theoretical and Applied Mechanics
of the Czech Academy of Sciences,
Prague, Czechia

Chapter 1

Vibration Control in Bridges

Kenneth C. Crawford

Abstract

The purpose of this chapter is to examine methods to control induced vibrations in steel and reinforced concrete (RC) highway bridges caused by three primary vibration forces, specifically wind, heavy traffic, and seismic events. These forces manifest their effects in bridge structural elements to different degrees, from small vibrations to large forces causing destruction of the bridge. This chapter examines bridge failures caused by induced vibrations, from wind loading, traffic loading, and seismic vibration loading and presents solutions developed to compensate for these vibrations. Bridge failures from seismic vibrations are the most destructive and are described in two major earthquakes in California. A major bridge failure from induced wind vibrations is considered, and two bridge failures caused by vibrations from heavy traffic loading are described. With lessons learned from these and other bridge failures, new design criteria and methods have been established to reduce and mitigate the destructive forces of induced vibrations. Significant changes in bridge structural engineering codes and design philosophy were made. While bridge structural design improvements have reduced the effects of wind, seismic, and heavy traffic vibrations, further research is needed to mitigate the long-term effects of vibrations on bridge performance and structural integrity.

Keywords: bridges, wind, heavy traffic, seismic events, vibrations, failures

1. Introduction

A national highway system is a large and extensive infrastructure made up of roads, tunnels, bridges, interchanges, ramps, and embankments in a complex combination of designs and configurations. The continuous and uninterrupted flow of traffic in a highway system is vital to a nation's economy and flow of goods and services. The efficient operation of a highway transportation system is dependent on its original construction and maintained condition of its key elements, in particular bridges. Steel and RC bridges are a critical component in a national highway system requiring sound design and high quality construction with an effective long-term inspection and maintenance program. The performance and survivability of bridges, under wind and heavy traffic loading, and in a natural disaster, such as a seismic event, is critical in a nation's transportation network. The performance of RC highway and steel bridges under these induced forces is a function of their ability to withstand damaging forces and induced vibrations in critical structural members. Many bridges constructed before the 1970s, both in Europe and the USA, do not meet current seismic design standards and are potentially subject to possible failure in the event of a major earthquake. Concrete and steel bridges constructed today are designed to perform to high standards under wind, heavy traffic, and seismic vibration loading.

1.1 Objective of chapter

Considering the impact of vibrational forces on bridge structures the objective of this chapter is to examine the nature of induced vibrations in bridges from natural and man-made forces and to gain an understanding on how these vibrations influence a bridge's structural performance. Induced vibrations from wind, heavy traffic and seismic events can significantly degrade the structural integrity of a bridge and its ability to sustain its designed load performance. To consider and understand the effects of induced vibrations in bridges this chapter examines several cases of bridge failures that have resulted from wind and heavy traffic loading, and from seismic events that destroyed a large number of bridges in a highway interstate system. The goal is to learn from these bridge failures and what role the induced vibrations played in their failure. The lessons learned in studying bridge failures provide an opportunity to better understand the forces of nature and the types of vibrations they induce in bridge structures and to be better able to develop improved bridge design codes, criteria, methods, materials, and construction process. A bridge today should not fail from wind, heavy traffic, or seismic induced vibrations.

2. Bridge failures from induced vibrations

While bridge failures occur for a number of reasons, normally through the deterioration of materials in structural members, the study of vibrations in a bridge's structural integrity provides an insight on how a bridge will perform under the stress of severe vibration loading. The effects of wind vibration are examined in two cases in which bridges failed from inadequate design. Two bridges are considered that failed from vibrations induced by heavy traffic loading. The third category of bridge failures is the result of destructive forces induced by the vibrations from seismic events. Earthquakes, depending on the magnitude of vibrations, are a major factor in the disruption and destruction, of highway bridge networks. The point in studying bridge failure mechanisms from induced vibrations is to learn and to establish better designs to improve long-term bridge performance to mitigate destructive vibrations forces.

2.1 Bridge failures from wind vibrations

Since the early 1800s when the first suspension bridges were being designed and built in England and the United States little was understood about wind induced vibrations. Vibrations from seismic events and heavy and traffic loading were not an issue at the time. After a number of wind induced suspension bridge failures in the first half of the 1800s, **Figure 1**, it was not until 1840 and 1849 serious consideration was given by bridge designer John Roebling (1806–1869) to design for wind loading in suspension bridges.

In the mid-1800s a number of the early suspension bridges collapsed for various reasons, some with loss of life. Poor-quality iron, and shortfalls in design and constructions were identified as contributory causes. In the aftermath of the collapses it was recommended structures should be periodically inspected and chains should be load-tested. Unfortunately these actions were not mandatory. Various

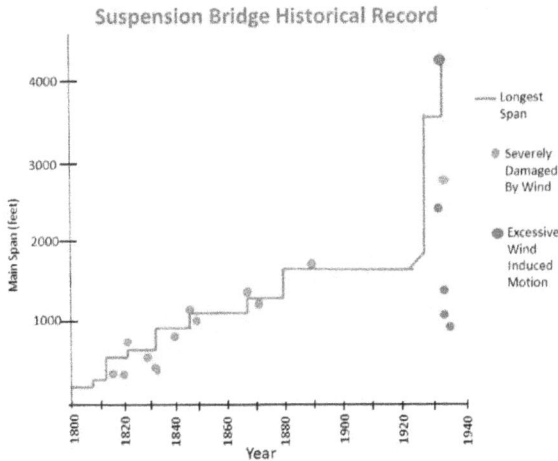

Figure 1.
Bridge failures from wind vibration loading in 1800s.

Figure 2.
Tacoma narrows bridge failure from a 40 mph wind.

design schemes were adopted to limit vibrations but the single most important was to stiffen decks by the addition of truss-parapets [1].

One of the classic examples of the effect of wind induced vibrations on a suspension bridge is the failure of the new Tacoma Narrows Bridge in Nov 1940, as the result of a 40-mile an hour wind, **Figure 2**.

Based on Austrian civil engineering deflection theory the engineer firm Moisseiff and Lienhard, New York City, who produced the original design for the bridge, stated the main cables were stiff enough to absorb wind pressure and stablilize the bridge, and assumed the wind forces would only push the bridge sideways. From the beginning the bridge had large vertical deflections even from moderate winds.

The primary explanation of the bridge failure was described as "torsional flutter." "Torsional flutter" is a complex mechanism. "Flutter" is a self-induced harmonic vibration pattern which can grow to very large vibrations. The external force of the wind alone was not sufficient to cause the severe twisting that led the Narrows Bridge to fail. It is noted the bridge deck's twisting motion caused torsional oscillation which became self-generating causing the bridge to absorbed more wind energy in a condition called "self-excited" motion [2].

While vortex shedding may occur in low wind speeds around 25–35 mph, and torsional flutter occurs at higher wind speeds of 80–100 mph, the instability in the Tacoma Narrows bridge caused by vortex shedding and torsional flutter occurred at

3

relatively low wind speeds, less than 25 mph. Because 0f the type of bridge design, and relatively weak resistance to torsional forces from the vortex shedding instability the bridge went into self-amplifying "torsional flutter", which is what destroyed the bridge [2].

2.2 Bridge failures from traffic load vibrations

It is rare that a bridge fails from the vibrations of heavy traffic loading but two cases stand out as examples of failure under heavy load vibrations. In each case the vibration loading exceeded the designed loading capacity of the bridge with the failures compounded by other factors, such as design flaws, construction deficiencies, inspection errors, and maintenance short falls.

2.2.1 I-35 W bridge collapse Minneapolis, MN

The 8-lane I-35 W bridge, known as bridge 9340, was constructed in 1967 and served over 140,000 vehicles per day. The bridge consisted of fourteen spans: nine spans were of steel multi-girder construction, two were of concrete slab construction, and the main three were of deck truss construction. The Minnesota Department of Transportation (MnDOT) was tasked with taking over annual bridge inspection beginning in 1993. Prior to this, it was federally inspected every other year. Inspection reports by MnDOT often indicated significant corrosion, rusting, warped plates, and other structural issues with the bridge. It was noted that the lack of redundancy in the main truss design meant that the bridge was vulnerable to a collapse if a single critical piece in the truss were to fail. Subsequent inspection reports expressed concern about the bridge's structural integrity, but no motion to close or drastically reinforce it was ever made [3].

In 1 August 2007, at 6 pm, the I-35 W Mississippi river bridge collapsed suddenly taking with it 111 vehicles, killing 13 people, and injuring 145. The bridge, **Figure 3**, was non-redundant and fracture critical, meaning if one member failed the entire bridge would collapse. Although the iron frame bridge with riveted gusset plates had supported heavy traffic volume for over 40 years it was a single half inch gusset plate, in a badly corroded condition, that failed along a line of rivets that caused the entire 250 foot bridge to fail. It was the additional weight of construction equipment plus the vibrations of rush hour traffic at the time that actually triggered the failure of the gusset plate [3].

Figure 3.
I-35 W Mississippi River Bridge collapse August 2007.

The deterioration of the gusset plates in periodic annual inspections was not labeled as potentially critical. While the gussets were identified as the root cause of this devastating collapse, the investigation found a combination of separate factors coming together led to the disaster: design flaws, inadequate inspection, MnDot policies not being followed, poor information flow, the organizational structure not addressing bridge conditions and safety. All these factors combined caused the bridge to collapse [3].

2.2.2 Hyatt Regency Hotel, Kansas City, MO Skywalk Collapse

In 1981 the Hyatt Regency Hotel in Kansas City, Missouri, suffered a structural failure of two of its three skywalks above the hotel atrium (**Figure 4**). At 7:05, July 17, with approximately 1600 people gathered in the hotel atrium for a tea dance, the fourth level walkway, suspended directly over the second-floor walkway, gave way and fell on the walkway below taking both walkways to the ground floor, killing 114 and injuring 216. The primary cause of the failure was the induced vibrations from a large number of people on the skywalks (overloaded) dancing to the rhythm of the music on the ground floor. It was the worst civil engineering failure in US history, since the collapse of Pemberton Mill over 120 years earlier. Many lessons and reforms for this structural failure contributed to engineering ethics and safety and to emergency management.

The root cause of the structural collapse was the failure of a hangar bolt bracket in fourth floor skywalk. Contributing factors: failure in engineer review of shop drawings of a field change in the skywalk hangar bolts, inadequate design of skywalks, and lack of oversight responsibility. Kansas City society was affected for years, with the collapse resulting in billions of dollars of insurance claims, legal investigations and city government reforms.

2.3 Bridge failures from seismic vibrations

Vibrations from seismic events have devastating effects on bridges and structures. Earthquakes in California and the bridge failures that resulted from large seismic vibrations are examined. The highway system in southern California is a complex network of Interstates, bridges, and flyovers, and is subject to significant deteriorating impacts from earthquakes and seismic induced vibrations.

The Interstate 5 (I-5) is the main Interstate highway on the West coast of the United States running largely parallel to the Pacific coast of the continental U.S. and

Figure 4.
Collapsed skywalk bridge in Hyatt Regency Hotel.

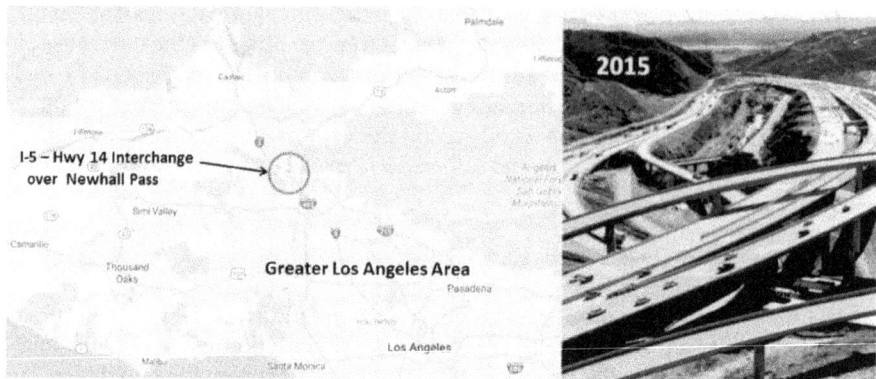

Figure 5.
Interstate I-5 and State Road 14 interchange over the Newhall Pass.

Route 99 from Mexico to Canada. The Golden State Freeway on I-5 begins one mile east of downtown Los Angeles and extends north through San Fernando Valley, across the Newhall Pass into the Santa Clarita Valley. I-5 then goes north from the Newhall Pass over the Grapevine Pass to eventually reach its second-highest point at Tejon Pass with an elevation of 1275 m, into the San Joaquin Valley, and further to Sacramento.

The Newhall Pass Interchange, **Figure 5**, is a major highway interchange north of Sylmar in Southern California, connecting Interstate 5 (Golden State Freeway) with State Route 14 (Antelope Valley Freeway) (SR 14). The interchange is extremely large, and consists of numerous flyover ramps and two tunnels. Portions of Interstate 5 in the pass reach up to 21 lanes wide. The complex interchange structure combines a directional T-interchange with a collector-distributor bypass.

2.4 Impact of seismic vibrations on California highway bridges

The failure of the Interstate 5 and SR14 at Newhall Pass interchange in southern California, along with other freeway overpasses, in the 1971 earthquake, and again in the Northridge earthquake in 1994, provide examples of the devastating impact seismic vibrations can have on the local economy and a critical state highway network in an urban area. The failure of highway bridges in these two earthquakes occurred because they did not meet the updated Caltrans bridge seismic design criteria and had not been retrofitted before the earthquakes occurred.

2.4.1 Sylmar earthquake bridge failures

The Sylmar earthquake (also known as San Fernando earthquake) took place near the San Fernando Valley in southern California on February 9, 1971. This earthquake was of magnitude 6.6 on the Richter scale and had an epicenter with coordinates of 34.41°N 118.40°W. The earthquake lasted 12 s and had a depth of 13 km (8.1 miles). Thrust faulting ruptured a segment of the San Fernando fault zone with a total surface rupture of 19 km with a maximum slip of 2 m. 65 persons died, 49 in the collapse of a Veterans Administration Hospital. 200 people were injured. An estimated $505–553 million occurred in structural damage.

In the Sylmar earthquake twelve overpass bridges failed and fell onto the freeways below. Major bridge failures occurred at the Interstate 5 and State Road (SR) 14 interchange, **Figure 6**. A total collapse of the southbound I-5 to northbound

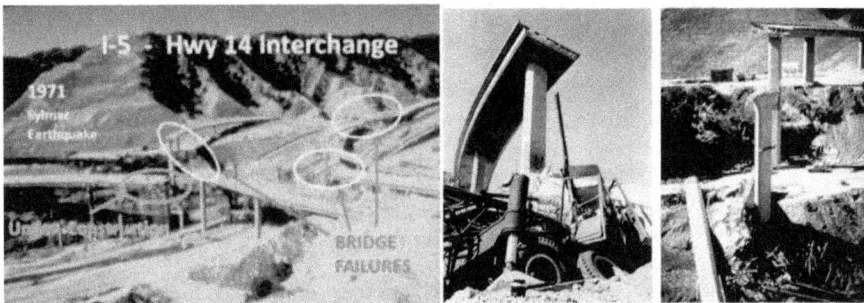

Figure 6.
1971 Sylmar Bridge Failures on I-5—State Road 14 interchange.

SR14 overpass occurred as a result of the earthquake. This collapse resulted in the additional collapse of the intersecting southbound SR 14 to southbound I-5 overpass (as this connector bridge was directly beneath the I5—SR 4 overpass) [4].

Both bridges fell directly onto the southbound I-5 truck bypass. There was damage to all bridge structures involved, varying from minor cracking and splaying, to the loss of complete sections of bridges. Most of the bridge damage in the Sylmar earthquake occurred on the I-5—SR14 interchange, **Figure 6**. Vibrations from this 6.6 magnitude earthquake caused the structure between two columns to separate from the actual supporting column which caused the highest overpass road (Newhall Pass) to collapse on top of the overpass below it, which then all collapsed onto the freeway below it. The rebuilt interchange was completed in 1973.

On the I-5—SR 14 interchange, it was noted the column that collapsed experienced damage at the ends, while the middle part of the column received little damage. Jennings noted, the small length of seating at the end of the fallen section, the lack of effective ties(steel reinforcing) to neighboring sections, and the general configuration of the inverted-pendulum structure were indicative of inadequate attention to the effects of strong earthquake motion. There are a variety of possible ways that the bridge structure might have failed, but two points are clear. First, the evidence strongly indicated a vibration failure. Permanent ground displacements (none were noted) were not thought to have played a significant role in the collapse [5].

2.4.2 1994 Northridge Earthquake Bridge Failures

The 1994 Earthquake (known as Northridge Earthquake) occurred on January 17, 1994 near Reseda, California, a neighborhood in the north-central San Fernando Valley region of Los Angeles. The epicenter was 34.213°N 118.537°W. The magnitude of the earthquake was 6.7 Mw (moment magnitude) and caused over $50 billion in damage. 57 people died and over 8700 were injured. 40,000 buildings were damaged and 20,000 were left homeless. While about $2 billion occurred in damaged transportation infrastructure (roads and bridges), over $50 billion in damages occurred with severe impacts on the local economy over the following 3 years. The lesson learned is that small damage to a transportation network can have a devastating impact on the local economy for years after [4].

The Northridge earthquake was the first earthquake since the 1933 Long Beach earthquake to strike beneath an urban area and occurred on a blind thrust fault producing the strongest ground motions ever recorded in urban areas in North America. Damage to freeways, office and apartment buildings, and parking structures was extensive. Structures were lifted of their foundations by high levels of

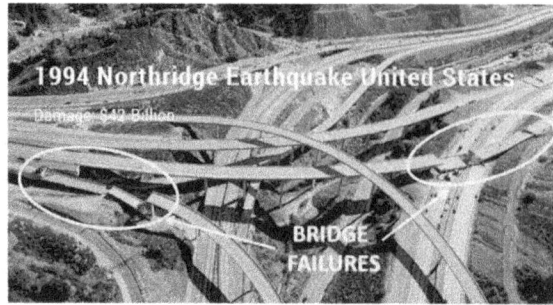

Figure 7.
Two bridge failures on I-5—State Road 14 interchange.

Figure 8.
I-5 Newhall bridge failure near SR14.

vertical and horizontal accelerations. Bridge damage was substantial. Over 4000 km^2 of the earth's crust deformed, forcing the land surface upward in the shape of an asymmetric dome [6].

The Northridge earthquake appears to be the result of a truncated fault that broke in the 1971 San Fernando earthquake at a depth of 8 km, resulting in a reverse skip of more than three meters along a 15-km long south-dipping thrust fault line. This fault raised the Santa Susana mountains by more than 70 cm. Because the fault was directly under the city the Northridge earthquake caused many time more damage. Of all the damage the biggest surprise was the fracture of welds in steel-frame buildings. Concealed thrust faults remain the greatest potential for strong ground shaking for Los Angeles, even in moderate quakes [7].

The Northridge earthquake collapsed seven freeway overpass bridges and caused the disruption of a large portion of the northwest Los Angeles freeway system, **Figures 7** and **8**. The Northridge earthquake caused the southbound SR14 to southbound I-5 connector to collapse, **Figure 6**, and a bridge crossing on the San Fernando Freeway.

2.4.2.1 Causes of Northridge seismic vibration structural failures

Over 350 square miles (900 square kilometers) received extensive damage to residential, commercial building, and regional lifelines from the main shock and aftershocks. Although the earthquake magnitude of 6.7 was moderate by earthquake standards, the neighboring communities of Sylmar, Burbank, Van Nuys,

Figure 9.
Failure of I-210 overpass to I-5 south.

Glendale, Santa Monica, Newhall, and San Fernando sustained damages. In looking at other earthquakes the 1971 San Fernando had a magnitude of 6.4, 1964 Alaska had 8.1, 1989 Loma Prieta 7.1, 1987 Whittier Narrows 5.9, and the 1991 Sierra Madre had a magnitude of 5.8.

There was also major bridge damage to the Golden State Freeway (I-5) and Foothill Freeway (I-210) Interchange, **Figure 9**. The westbound I-210 to southbound I-5, under construction and complete except for paving at the ramp section, collapsed over the I-5. Possible cause of failure was vibration that moved the overpass off its supports due to an inadequate column seat. Unlike the situation at the I-5—SR 14 Interchange, permanent ground movement (defined as several inches of left-lateral displacement with possibly an element of thrusting) was observed in the area.

Examination of the earthquake magnitude and epicenter and resulting bridge failure was made with consideration for the specific modes of bridge failure. Of the seven major freeway bridges that failed, two bridges on the SR 118 had flexure/shear failure of short and stiff columns and a low transverse reinforcement ratio. The I-5 bridge failure was from skewed geometry and unseating of expansion joints. The I-10 bridge failures were caused by flexure/shear failure of short stiff columns and brittle shear failure of stiff columns. The I-5—SR 14 bridge failures were caused by short column brittle shear failure. Five of the failed bridges were scheduled for retrofit which had not been accomplished before the earthquake. The two other failed bridges had been identified as not requiring retrofit [6, 8].

Figure 10.
Failed highway bridge columns on Santa Monica Freeway.

Figure 11.
Failed bridge columns on Santa Monica Freeway and I-5.

Figure 12.
Column failure of I-210 overpass to I-5 south.

2.4.2.2 Northridge bridge column failures

The Northridge earthquake caused extensive damage to bridge columns, particularly on the Santa Monica Freeway (I-10) from the I-5 to Santa Monica on the ocean and SR 118 further north, **Figures 10–12**. The column failures were caused by flexure and shear failures of short stiff columns and by the brittle shear failure of stiff columns, **Figures 10** and **11**.

The Northridge earthquake's high vertical acceleration was the primary force causing the bridge column failures. The rate of vertical acceleration was the highest ever recorded in North America. In studying the ground wave motion there was evidence of surface wave amplification which increased structural damage. The resulting seismic induced vibrations were devastating to the bridge columns not designed for the magnitude of the vibrations, resulting in widespread column destruction.

3. Lessons in controlling bridge wind vibrations

The construction of a suspension bridge consists of main towers, main suspension cables, anchorages for the cables, trusses, and stiffening girders. After the main cables are suspended between the towers and connected to their anchorages,

John Roebling's Suspension Bridges

1844 Allegheny Aqueduct – Pittsburg

1845 Smithfield Street Bridge - Pittsburg

1849 Delaware and Hudson Aqueduct - PA to NY

1855 Niagara Suspension Bridge

1856 Ohio River Bridge – Cincinnati, OH

1860 Sixth Street Bridge – Pittsburg

1883 Brooklyn Bridge - NYC

Figure 13.
Bridges designed and built by John Roebling.

vertical suspenders are connected to the deck which carries the traffic load. The main cables, made of high-strength steel, are the primary load carrying members of the bridge and are efficient in reducing and mitigating wind vibrations. Because the bridge structure deadweight is reduced longer suspension bridge spans are possible.

In the early to mid-1800s suspension bridges were most often light spans with flexible decks and were vulnerable to the aerodynamic forces of cross winds. To counter these wind forces bridge engineers began moving to heavier and stiffer suspension bridges. An example is the Brooklyn Bridge designed in 1883 by John Roebling to which he added mass and stiffness to resist the strong winds on the East River (**Figure 13**). However by the early 20th century bridge engineers failed to learn from past bridge failures. David Billington, Gordon Y.S. Wu Professor of Civil Engineering at Princeton University, stated "Roebling's historical perspective seemed to have been replaced by a visual preference unrelated to structural engineering, as seen in the failure of the Tacoma Narrows bridge" [2].

John Roebling recognized the problem of wind loading on suspension bridges and designed a number of features in the bridge to reduce the effects of wind vibrations. This involved adding stays and trusses to counteract wind loading.

Figure 14.
822 ft Niagara railroad suspension bridge.

Figure 15.
John A. Roebling Bridge Cincinnati OH.

These features were added to the 1849 Niagara railroad bridge and the 1855 Niagara suspension bridge.

The Niagara railroad bridge, **Figure 14**, opened in 1849, had a top deck for streetcars and a lower deck for pedestrians and wagons. Two tracks for streetcars were laid. Diagonal stays were added to increase load capacity, strengthen the floor, and check train and wind vibrations. Wrought iron trusses were added, running the length of the bridge.

With the design and construction of the Cincinnati river bridge, **Figure 15**, opened in 1867, additional features were added to mitigate the effects of wind. Roebling made further improvements for wind vibration stability by adding additional stays and trusses and increasing the mass of the bridge. These features were then used for the design of the Brooklyn Bridge in which stiffening girders and additional mass were added to provide a stable bridge deck for the high winds on the East River. With the design of the Brooklyn Bridge Roebling had a good understanding of the relationship of suspension bridge span length and the forces of induced wind vibrations. This understanding would not extend to the suspension bridges designed and built in the next century.

After a series of suspension bridge failures in the 1800s and following the Tacoma Narrows Bridge failure, professor J.K. Finch, Columbia University, civil engineering, in an Engineering New-Record article, stated: "These long-forgotten difficulties with early suspension bridges clearly show that while to modern engineers, the gyrations of the Tacoma bridge constituted something entirely new and strange, they were not new — they had simply been forgotten. An entire generation of suspension-bridge designer-engineers forgot the lessons of the 19th century." Prior to the Tacoma Narrows bridge failure the last major suspension bridge failure was the Niagara-Clifton Bridge, constructed in 1847, which collapsed in a storm in 1889. Well into the 1930s, aerodynamic forces on bridges were not well understood and researched, resulting in insufficient wind loading designs for suspension bridges [2].

4. Lessons controlling heavy load bridge vibrations

The bridge failure cases in Section 2.2 illustrate what the impact of uncontrolled vibrations can have on bridge structural integrity. Through extensive research new methods and devices have been developed to dampen and significantly reduce vehicle (truck, train) induced vibrations in bridges. One such device is the tuned mass damper (TMD), also known as a seismic damper of harmonic absorber.

Consisting of a mass mounted on damping springs the TMD is mounted on bridges and buildings to reduce mechanical vibrations. The TMD's oscillation frequency is tuned to synchronize with the resonant frequency of the object to which it is mounted.

Tuned mass dampers stabilize against violent motion caused by harmonic vibration. They use a comparatively lightweight component to reduce the vibration of a system so that its worst-case vibrations are less intense. TMDs can prevent damage and structural failure and are frequently used on highway and rail bridges to reduce harmonic vibrations [9].

Using the modal properties of a bridge structure in bridge engineering the TMD is designed to reduced vibrations in bridges. Depending on the design methodology of the TMD, bridge vibrations in certain frequencies are reduced while other frequencies are amplified.

One application of the TMD is on railroad bridges. A moving mass model is used to consider the dynamic response of the bridge to the moving train load. Zhaowei Chen et al., introduce the design methodology of bridge-based designed TMD (BBD-TMD) in which a detailed train-track-bridge coupled dynamic model with attached BBD-TMD is established based on the multi-body dynamics theory and the finite element method. To evaluate the running performance of a train, three indicators are selected, namely wheel-axle lateral force (Fw), derailment coefficient (DC), and wheel unloading rate (WUR). The authors note the indicator WUR is aggravated in some cases by the BBD-TMD, indicating that the performance of a running train at different speeds should be seriously considered in designing TMDs based on the bridge modal property [10].

Using the train wheel and body for a 2-degree of freedom (DOF) system a high-speed train such as the TGV can be modeled. For a three-span bridge using the midpoint vertical displacements and the fast Fourier transform and comparing the results before and after installation, the efficiency of the TMD can be shown. The TMD has a variety of merits in that it has permanent service time, and only requires easy management and maintenance efforts and no external power supplying sources [10].

5. Lessons learned for controlling seismic vibrations in structures

No single earthquake like the 1994 Northridge earthquake had such an impact on the art and practice of structural engineering, causing an overall and widespread reevaluation of engineering practices so deeply rooted in structural engineering. Following 1994 the most significant changes by the Seismic Advisory Board (SAB) in the philosophy and design practice of structural engineering were in the following areas:

a. Understanding the dire consequences of unjustified extrapolations.

b. Significant improvements in knowledge of design characteristics of strong ground motions.

c. Limits on the understanding of ultimate behavior of structures.

d. Value of seismic instrumentation and its widespread use.

e. Insufficiency of design practice targeted only to life-safety and collapse prevention.

f. Development and application of performance based design methodologies.

g. Significant changes and additions to building codes and standards.

Prior to the 1994 Northridge earthquake welded steel moment resistant frames (WSMRF) in steel-frame buildings were considered the most reliable and sought after structural systems for earthquake-resistant construction. The poor performance of WSMRFs in the Northridge earthquake, with failures in steel frame welds because of the ductility of steel with ground movement forces, revealed the limitations on the valid bounds of experimental research on which the use of WSMFR was based [1]. The failures were in extrapolations of limited steel frame seismic tests on which designs were based and in the failure of the processes and quality of steel welding [8].

The lesson learned in the Northridge earthquake provided key lessons learned for earthquakes that occur beneath cities, indicating the concealed faults under Los Angeles are far more complex than previously thought. The one primary lesson learned shows urban areas can be subjected to ground motions with peak accelerations approaching the force of gravity, exceeding the levels of shaking anticipated by building codes. It is essential building codes address the forces expected on buildings subject to earthquakes.

The US Geological Survey (USGS) has established national and regional maps of probabilistic earthquake ground shaking through the USGS (US Geological Survey). The National Seismic Hazard Mapping Project (NSHMP) integrates the results of research in historical seismicity, paleo-seismology, strong motion seismology, and site response taking into account all the possible locations and magnitudes that are likely to happen in future hypothetical earthquakes. By the year 2000, all US model building codes incorporate ground motion hazard maps derived from the USGS studies to insure structures are engineered to have the appropriate level of resistance to earthquake ground motion [8, 9].

5.1 Improvements in highway bridge seismic design

The development of seismic design criteria and guidelines in the US, and particularly in California, has evolved in response to substantial earthquake damages to bridges and structures. Following the 1971 Sylmar earthquake near Los Angeles a major bridge retrofit program was started by the California Department of Transportation Caltrans). Following the devastating 1989 Loma Prieta earthquake in Oakland, CA, the Governor of California created, in 1990, the Seismic Advisory Board (SAB) under Caltrans. The SAB is an independent body whose role is to advise Caltrans on seismic policy and technical practices to enhance the seismic safety and functionality of California's transportation structures. The SAB initiated significant changes in bridge and structural design philosophy and criteria, moving from established prescriptive criteria to developing performance-based seismic engineering (PBSE) concepts and methodologies. FEMA-273 guidelines (1997) made PBSE a reality for structural engineers by defining performance of components and systems in terms of a spectrum varying from continuous operation to collapse prevention [11].

5.2 Bridge seismic vibration response

The response of RC highway bridges under conditions of a natural disaster such as an earthquake is a function of its design and construction and its ability to withstand the vibrations and energy of local ground shaking without collapse. In analyzing the ground motion of the 1994 Northridge earthquake, the strength of ground shaking was measured in the velocity of ground motion, the acceleration of ground motion, the frequency content of the shaking and duration of the shaking. When assessing the potential shaking hazard at a site where frequent strong motion is expected to re-occur, consideration is made of the characteristics of waves produced by an earthquake rupture, also strongly influenced by the fault rupture orientation, its depth, and the details of how the slip spread across the ruptured fault patch. In the Northridge earthquake the rupture of a concealed fault beneath an urban area caused widespread and extensive damage to bridges and buildings [12].

5.3 Bridge seismic vibration design for critical components

The ability of a RC highway bridge to withstand damaging vibrations from an earthquake depends on seismic design guides available at the time of the bridge design and construction. Bridges designed and constructed before the 1971 Sylmar earthquake did not comply with the latest seismic design guides. Following three relatively moderate earthquakes between 1971 and 1994, two in southern California and one in Oakland, California, which resulted in significant infrastructure damage, Caltrans, led by the Seismic Advisory Board, made a substantial effort to redefine and improve their seismic structural codes.

In their Seismic Design Criteria, Caltrans defines a ductile member in an RC bridge as any member that is intentionally designed to deform in elastically for several cycles without significant degradation of strength or stiffness under the demands generated by the Design Seismic Hazards. The Design Criteria states seismic-critical members may sustain damage during a seismic event without leading to structural collapse or loss of structural integrity. Bridge components are designated as seismic-critical if they will experience any seismic damage as determined by the project engineer and approved during type selection. Ductile and seismic-critical members are defined as columns, Type I shafts, pile/shaft groups and Type II shafts in soft or liquefiable soils, pier walls, and pile/pile-extensions in slab bridges (designed and detailed to behave in a ductile manner). Other bridge components such as dropped bent cap beams, outrigger bent cap beams, "C" bent cap beams, and abutment diaphragm walls are to be designed as seismic critical. All other components not seismic-critical shall be designed to remain elastic in a seismic event [13, 14].

5.4 Seismic vibration control in buildings and bridges

Seismic vibration control consists of technologies to reduce the seismic effects in structures (building and bridges), and thus minimize earthquake damage. When the seismic waves travel upwards through the base of buildings and bridges, reflections considerably reduce the wave energy.

To control residual energy in seismic waves, which are the source of major damage in earthquakes, seismic vibration control technologies are used with dampers, which absorb energy over a wide range seismic wave frequencies. Seismic energy flow into buildings is also controlled by isolating the building with pads mounted in the base load carrying elements decoupling the building superstructure from the foundation substructure.

An excellent example of bridge engineering to control seismic and wind vibrations is provided by the Rio Antirrio bridge, **Figure 16**, over the Gulf of Cornith in Greece.

To absorb the energy from earthquakes it was necessary to develop new design and construction techniques for the piers and pier footings. The pier footings were

Figure 16.
2380 m Rio Antirrio fanned cable-stay bridge.

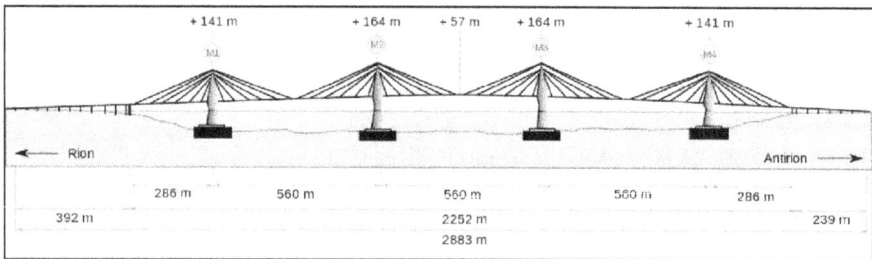

Figure 17.
Rio Antirrio bridge with four pylons and five spans.

not buried in the sea bed but mounted on top of a gravel bed allowing the piers to move laterally while the gavel bed absorbs the earthquake energy. To absorb movement in the bridge deck jacks and dampers were connected to the bridge pylons. Protection from the effect of high winds on the decking is provided by the use of aerodynamic spoiler-like fairing and on the cables by the use of spiral Scruton strakes.

The design of the pylon footings, not anchored in the seabed, but simply placed on a level bed of rock is unique in structural engineering for controlling seismic vibration (**Figure 17**). The bed of rock absorbs the energy of seismic vibrations without transmitting the energy through the pylons to the bridge structure. The bridge is designed to withstand an earthquake up to a 7.4 magnitude.

6. Role of FRP systems for bridge stability in seismic vibrations

Since the mid-1990s advanced composites were first used In the United States for seismic retrofit of buildings and bridges which have also seen a significant increase worldwide in the use of advanced composite materials for bridge rehabilitation to repair damaged structures, to strengthen structures for increased demand (vehicle loading), and to retrofit structures to control seismic vibrations. Referred to as fiber reinforced polymers (FRPs), these advanced composite materials consist of glass, carbon, or aramid fibers embedded in a polymer matrix. These FRP overlays successfully used to strengthen columns and girders for shear, reinforced concrete slabs in flexure, and on joints for external cap and column connections. During the Caltrans Phase I and Phase II Bridge Column Retrofit Program, early applications of

FRPs were developed and implemented demonstrating significant benefits over the more conventional steel jacketing.

Other key benefits of FRPs are their high mechanical characteristics and their light weight (up to 5 times stronger and 5 times lighter than mild steel). Other issues the Seismic Advisory Board believes need to be better addressed for broad based FRP applications concern fire resistance and quality control/inspection measures during and after the retrofit installation. For FRP applications to be used on California bridges and to meet life safety codes Caltrans requires testing for proof of concept and performance validation testing of FRP technology for bridge retrofit. Although the retrofit concept was tested on scaled bridge components, final proof testing was done on full scale bridge elements [8].

6.1 Example of FRP bridge retrofitting

To accomplish stability in bridge structures from seismic vibrations it is necessary to strengthen and stabilize bridge bents and foundations for seismic loading. Analysis considers types of retrofitting applications on bridge bents and the advantages of FRP systems to increase structural member axial capacity and ductility in columns and beams. Retrofitting RC bridges by encasing and strengthening bridge bents with CFRP (carbon fiber-reinforced polymers) systems provides an effective method of mitigating the impacts of seismic vibration loading. The carbon fiber and epoxy resin composite for carbon-fiber reinforced polymer material (CFRP) has 28,000 unidirectional carbo fibers per tow with 6.5 tons per 25.4 mm, a modulus of elasticity of 65 GPa, a tensile strength of 628 MPa, ultimate axial strain of 10 mm/m, and a layer thickness of 1.32 mm.

For CFRP retrofitting of bridge bents, **Figure 18**, it is necessary to develop a higher base shear and moment capacity than the existing foundation and pile cap system. Performance-based design determines column CFRP jacket thickness for plastic hinge confinement, shear strengthening, and lap splice clamping. The design increases displacement ductility of the bridge bent developing a higher base shear and moment capacity. The RC beam connecting the pole cap completes the tension and compression load path, increasing shear and flexural capacity of the foundation [15, 16].

Many studies on the use of FRP jackets 0n reinforced concrete (RC) columns have shown the jackets are effective in increasing shear capacity and flexural durability in the columns. However the contribution of FRP jackets for flexural strength for small axial loads is minimal. While applying FRP sheets in the direction of a column is difficult because difficulties with base anchorage, a research project

Figure 18.
CFRP strengthening of bridge bents and columns.

to upgrade the flexural capacity of RC piers used near-surface mounted (NSM) FRP rods. Flexural strengthening was achieved using NSM carbon FRP rods anchored into the footings. The piers were tested under static push/pull load cycles [17].

Another test was performed on RC piers in which three of the four piers were configured with different combinations of FRP rods and jackets. Using an analytical model with given load levels the net forces acting on the bridge pier were determined with strengthening techniques and modes of failure and confirming the effectiveness of the technology to strengthen RC piers [17].

The performance of highway bridges in earthquakes over the past several decades has been less than satisfactory because of poorly designed details and outdated design principles. In an effort to improve bridge performance in seismic events tests were conducted on the South Temple Bridge, built in 1963, during the I-15 reconstruction project in Salt Lake City. Five reinforced concrete bridge bents were tested, with three bents in as-is condition, two bents after a carbon fiber reinforced polymer (FRP) composite seismic retrofit, and one bent after a carbon FRP composite repair. The lessons learned from these tests were used in developing improved recommendations for the seismic retrofit design of bridge T-joints using FRP jackets. Using a nonlinear pushover static analysis of the as-is bent the performance-based design procedure includes determination of the column FRP jacket thickness for plastic hinge confinement, shear strengthening, and lap splice clamping. Using three elements the FRP jacket in the T-joints consists of diagonal FRP composite sheets for resisting diagonal tension, FRP composite sheets in the direction of the beam cap axis for shear strengthening and increased flexural capacity, and U-straps clamped at the column faces that go over the beam cap. The in-situ tests demonstrated that application of an external FRP composite seismic retrofit to concrete bridges with inferior seismic design details provides adequate ductility and seismic performance [17].

In a 2001 project in the Republic of Macedonia 19 slab and girder highway bridges were strengthened with CFRP plates to strengthen the bridges to NATO military load class 100 to compensate for induced vibrations from heavy military transports. The CFRP strengthening increased the bending moment and load capacity of the bridges by 60%. In 2019, non-destructive testing (NDT) was conducted on 12 of the 19 bridges to determine the condition of the CFRP plate-concrete bond. The results of the NDT field survey indicated 100% of the CFRP plates remain bonded to the bridge structural members 18 years after application. The use of CFRP material is a proven technology to reduce heavy traffic vibrations on RC highway bridges [18, 19].

7. Conclusion

The purpose of this chapter on vibration control is to examine the effects of wind, heavy traffic, and seismic vibrations on steel and RC highway bridges by providing an overview of bridge performance and failure under different types of induced vibration loading. Cases of bridge performance are presented for wind, heavy traffic, and seismic vibration loading. Bridge failures from induced vibrations are presented with analysis of why bridges failed. The effects of seismic vibrations on RC highway bridges, in particular from earthquakes in California, with extensive destruction of highway bridge networks, led to changes in bridge design. The lessons learned following extensive bridge damages from the 1971 Sylmar, 1989 Loma Prieta, and 1994 Northridge earthquakes, causing major damage and disruptions in the economy and highway systems in southern California, drove the California Seismic Advisory Board (SAB) to make significant changes in design philosophy in structural engineering and the approach Caltrans uses in designing and constructing bridges. The major change was moving from established prescriptive criteria to developing performance-based seismic engineering (PBSE) concepts and methodologies. These changes have become standard in national structural engineering practices. Lessons learned from bridge failures from wind and heavy traffic loading vibrations have led to improved design codes and criteria, better construction methods, and improved materials and processes. While modern bridge design addresses vibrations for heavy traffic loading, the natural forces of wind and seismic events will always be a challenge for bridge engineers. The study and analysis of past bridge failures from induced vibrations provides lessons on how a bridge's structural configuration responds to the destructive forces of vibrations. Bridge engineers have made significant progress in designing for these forces. Today's highway and rail bridges meet high structural standards. The challenge going forward is to continue to improve bridge designs and construction to meet ever changing vibrational forces.

Conflict of interest

The author declares he has no conflict of interest.

Author details

Kenneth C. Crawford
Institute for Bridge Reinforcement and Rehabilitation (IBRR),
Bloomington, IN, USA

*Address all correspondence to: ken.crawford@ibrr.org

IntechOpen

References

[1] Tilly G. Dynamic behavior and collapses of early suspension bridges. Bridge Engineering. 2011;**164**(2):75-80. DOI: 10.1680/bren.2011.164.2.75

[2] APS News, November 7, 1940: Collapse of the Tacoma Narrows Bridge. This Month in Physics History. 2013;**14**(10)

[3] Crawford K. Elements of bridge deterioration and failure. In: 13th Intl Conference on Short and Medium Span Bridges; Toronto. Surrey, British Columbia: Canadian Society of civil Engineers; July 2022

[4] SCVTV. I-5 Freeway Overpass—1994 Northridge Earthquake. California: Santa Clarita; 1994. Available from: http://scvhistory.com/scvhistory/sc9402.htm

[5] Jennings PC. Engineering Features of San Fernando Earthquake in 1971. Pasadena, CA, USA: CalTech; 1971. Available from: http://authors.library.caltech.edu/26440/1/7102.pdf

[6] Finn WD, Ventura CE, Schuster ND. Ground motions during the 1994 Northridge earthquake. Canadian Journal of Civil Engineering. 1995;**22**(2):300-315

[7] FHA (Federal Highway Administration). The Northridge Earthquake: Progress Made, Lessons Learned in Seismic Resistant Bridge Design, Public Roads – Summer 1994. Vol. 58(1). Pasadena, CA USA: California Institute of Technology (CalTech); 1994. pp. 300-315

[8] Seismic Advisory Board (SAB). The Race to Seismic Safety. San Francisco, CA, USA: CalTrans (California Department of Transportation); 2003. pp. 100-102

[9] Ho-Chul WL. Vibration control of bridges under moving loads. Computers and Structures. 1998;**66**(4):473-480

[10] Chen Z, Fang H, Han Z. Influence of bridge-based designed TMD on running trains. Journal of Vibration and Control. 2018;**25**(1):182-193. DOI: 10.1177/1077546318773022

[11] Naeim F. The Impact of the 1994 Northridge Earthquake on the Art and Practice of Structural Engineering. Los Angeles, CA, USA: John A. Martin & Associates, Paper; 2004

[12] Boore DM et al. Estimated ground motion from the 1994 Northridge, California, earthquake. Bulletin of Seismological Society of America. 2003;**93**(6):2737-2751

[13] Seismic Advisory Board (SAB). The Continuing Challenge: The Northridge Earthquake of January 17, 1994. California Department of Transportation's (CalTrans). Vol. 58, Issues 1. Washington DC, USA: Federal Highway Administration Magazine "Public Roads"; 1994

[14] California Department of Transportation (CalTrans). Seismic Design Criteria (SDC). 2013. Ver. 1.7

[15] Crawford K, Krsteski S. FRP-retrofitting RC bridges for stabilization in earthquakes, lupine publishers. Trends in Civil Engineering and Architecture. 2018;**1**(4). DOI: 10.32474/tceia.2018.01.000120

[16] Pantelides C et al. Design of FRP Jackets for seismic strengthening of bridge T-joints. In: 13th Conference on Earthquake Engineering; Vancouver, B.C., Canada. Vancouver, British Columbia: 13 WCEE Secretariat; 2004. Paper 3127

[17] Alkhrdaji T, Nanni A. Flexural strengthening of bridge Piers using FRP composites. Structures Congress. ASCE Advance Technologies in Structural

Engineering. 2000;**200**:1-13.
DOI: 10.1061/40492(2000)174

[18] Nikolovski T, Dimitrievski T, et al.
Basic principles for the design of
strengthening using CFRP systems. In:
9th Symposium of MASE Proceedings;
Ohrid, Macedonia. Vol. I. Skopje, North
Macedonia: Macedonian Association of
Structural Engineers; 2001. pp. 48-54

[19] Crawford K. NDT evaluation of
long-term durability of CFRP-structural
systems applied to RC highway bridges.
International Journal of Advanced
Structural Engineering (IJASE).
2016;**8**:161-168

Fractional Optimal Control Problem of Parabolic Bilinear Systems with Bounded Controls

Abella El Kabouss and El Hassan Zerrik

Abstract

The purpose of this paper is to study a fractional distributed optimal control for a class of infinite-dimensional parabolic bilinear systems evolving on a spatial domain Ω by distributed controls depending on the control operator. Using the Fréchet differentiability, we prove the existence of an optimal control depending on both time and space, that minimizes a quadratic functional which leads into account, the deviation between the desired state and the reached one. Then, we show characterizations of an optimal distributed control for different admissible controls set. Moreover, we developed an algorithm and give simulations that successfully illustrate the theoretically obtained results.

Keywords: infinite-dimensional system, parabolic bilinear systems, fractional derivative, optimal control

1. Introduction

In engineering and mathematics, control theory deals with the behavior of dynamical systems. The desired output of a system is called the reference. When one or more output variables of a system need to follow a certain reference over time, a controller manipulates the inputs to a system to obtain the desired effect on the output of the system, As an example: the control of vibration which is becoming more and more important for many industries. This generally has to be achieved without additional cost, and thus, detailed knowledge of structural dynamics is required together with familiarity of standard vibration control techniques. We also cited the following works on what concerns the vibration control [1–3].

The bilinear system involves the product of state and control, linear in state and linear in control but not jointly linear in state and control. The interest of these systems lies in the fact that many natural and industrial processes have intrinsically bilinear structures, This is the case of furnaces for heating metal slabs or heat exchangers, aircraft and robot arms, or energy transmission lines.

Let Ω be an open bounded domain of $\mathscr{R}^n$, $n \geq 1$, with regular boundary $\partial\Omega$, and consider a bilinear system described by the equation (see [4])

$$
\begin{cases}
\dfrac{dz}{dt}(x,t) = Az(x,t) + u(x,t)Bz(x,t) & Q = \Omega \times]0, T[, \\
z(x,t) = 0 & \Gamma = \partial\Omega \times]0, T[, z(x,0) = z_0(x)\Omega,
\end{cases}
\tag{1}
$$

IntechOpen

where, $A = \Delta$ of the domain $\mathcal{D}(A) = H_0^1(\Omega) \cap H^2(\Omega)$, u is a control assumed to belong to the set of controls

$$U = \{u \in L^2(Q)/ - m \leq u \leq M\} (with \quad M \geq m > 0). \tag{2}$$

B is a bounded control operator on $L^2(\Omega)$. For $z_0 \in H_0^1(\Omega)$ and $u \in U$, system (2) has a unique solution $z \in W = \{z \in L^2(0, T; H_0^1(\Omega)) | \frac{\partial z}{\partial t} \in L^2(0, T; L^2(\Omega))\}$.

Let us consider the fractional quadratic control problem:

$$J(u^*) = \min_{u \in U} J(u), \tag{3}$$

with

$$J(u) = \frac{1}{2} \|D_x^\alpha z - z_d\|_{L^2(0,T;L^2(\Omega))}^2 + \frac{\beta}{2} \|u\|_{L^2(0,T;L^2(\Omega))}^2, \tag{4}$$

where, D_x^α denotes the fractional spacial derivative of order $\alpha \in]0, 1[$, z is a solution of system (2), $z_d \in L^2(\Omega)$ is a desired derivative and β is a positive constant.

Ractional calculus has emerged as a powerful and efficient mathematical instrument during the past six decades, mainly due to its demonstrated applications in numerous, seemingly diverse, and widespread fields of science and engineering. As an example, The theory of fractional differential equations has received much attention, as they are important for describing the natural models as in diffusion processes, stochastic processes, economics, and hydrology. Moreover, the fractional optimal control has been studied in many works, such as Frederico et al. have studied a fractional optimal control problem in Caputo's sense. Agrawal [5] have presented an extended approach to a class of distributed system whose dynamics are defined in the sense of Caputo. In [6], they considered the fractional optimal control problem for variable inequalities. In [7], Bahaa studied the fractional optimal control problem for different systems. When $\alpha = 0$, problem (2) was considered in many works.: Bradley and Lenhart [8] have shown the existence of such an optimal control and given characterization of such control using necessary optimality conditions. Then, an optimal distributed control for a Kirchhoff plate equation acting on the state position. Also, they collaborated with Yong [9] on the same equation by temporal controls acting on the speed state and with special optimal control in Bradley and Lenhart [10]. For parabolic systems, we have mentioned the work in [11], which established an optimal control of a parabolic equation, modeling one-dimensional fluid through a soil-packed tube in which a contaminant is initially distributed, taking a functional criterion as a combination of the final amount of contaminant and the energy. In the same way, Addou and Benbrik [12] studied a fourth-order parabolic distributed parameter system and derived the existence and uniqueness of temporal bilinear optimal control. Then, Zerrik and El Kabouss [13] extended this problem to a more general class of systems governed by a fourth-order parabolic operator and excited by bounded and unbounded controls. A wide literature has also been considered for infinite hyperbolic systems, especially, by Liang [14] who analyzed an optimal control problem for a wave equation with internal bilinear control, and has given an optimal control that allows minimizing a functional cost which contains the difference between the solution's position and a desired one. In the case of boundary bilinear controls: Lenhart and Wilson [15] have studied the problem of controlling the solution of the heat equation with the convective boundary condition, such as, that the bilinear control

represents a heat transfer coefficient. The used approach consists in finding a
unique optimal control in terms of the solution of an optimality system.

For a system evolving on a spatial domain Ω, regional controllability concerns
the extension of the classical notion of controllability (controllability on the whole
domain Ω) to the controllability only on a subregion ω of Ω. This notion is
interesting for many reasons: it is close to real applications. For instance, the physical problem that concerns a tunnel furnace where one has to maintain a prescribed
temperature only in a subregion of the furnace and may be of great help for systems
that are non-controllable on the whole domain but controllable on some subregions,
and controlling a system on a subregion $\omega \subset \Omega$ is cheaper than controlling it in the
whole domain. Zerrik and El Kabouss [16] have studied a regional optimal bilinear
control of wave equation, taking a functional cost as the sum of the energy and the
difference between the solution of the wave equation and the desired state for
bounded and unbounded controls. Recently, Zerrik and El Kabouss [17] established
an output optimal control problem with a bounded control set. In other words, they
considered a problem of controlling only an output of the solution of a parabolic
system. In [18], they have studied an optimal control problem for the heat equation
in order to give control that leads to a state as the class as possible to the desired
state, only on a subregion of the domain of evolution, under constrained
controls sets.

In this paper, we consider $0 < \alpha < 1$, which is very important for modeling many
real processes. We study a fractional optimal control problem of parabolic bilinear
systems. Using the Frechet differentiability, we prove the existence and give the
expression of an optimal control solution of (2). Then we discuss particular cases of
admissible controls set.

2. Existence of an optimal control

This section discusses the existence of a solution of the problem (2).
First, let us recall the notion of the weak solution of the system (2).
Definition 1.1.
Let $T > 0$, a continuous function $z \in [0, T] \rightarrow L^2(\Omega)$ is a weak solution of system
(3) on $[0, T]$, if it satisfies the following integral equation

$$z_u(t) = S(t)z_0 + \int_0^T S(t - s)u(., s)z(s)ds, \quad \text{for all} \quad t \in [0, T] \tag{5}$$

where $S(t)$ denotes the C_0 semi-group generated by A in $L^2(\Omega)$.
For fractional Riemann Louiville derivatives, we recall the following definition.
Definition 1.2.
Let $0 < \alpha < 1$ and $T > 0$, the fractional spatial Riemann Liouville derivatives of
order α is defined by:

$$D_x^\alpha : H_0^1(\Omega) \rightarrow L^2(\Omega) \tag{6}$$

$$z \rightarrow D_x^\alpha z = \frac{d}{dx} I_0^{1-\alpha} z, \tag{7}$$

where $I_0^{1-\alpha}$ is the Riemann-Liouville integral of $(1 - \alpha)$ order defined by:

$$I_0^{1-\alpha} z(x, t) = \frac{1}{\Gamma(1 - \alpha)} \int_0^x (x - \tau)^{-\alpha} z(\tau, t)d\tau \tag{8}$$

with $\Gamma(1 - \alpha) = \int_0^{+\infty} \tau^{-\alpha} e^{-\tau} d\tau$.

In the following, we show the existence of optimal control, solution of problem (3).

Theorem 1.3.

Problem (3) has at least one solution.

Proof: For $u \in U$, the associated solution of system (3) is one of the equation

$$z_u(x,t) = S(t)z_0(x) + \int_0^T S(t - s)u(x,s)Bz(x,s)ds. \tag{9}$$

Using the bound of the semi-group $(S(t))_{t \geq 0}$ over $[0, T]$, we have

$$\|z_u(t)\|_{L^2(\Omega)} \leq C\|z_0\|_{L^2(\Omega)} + C\|B\|_{L^2(\Omega)} \int_0^T \|u(s)z(s)\|_{L^2(\Omega)} ds. \tag{10}$$

It follows

$$\|z_u(t)\|_{L^2(\Omega)} \leq C\|z_0\|_{L^2(\Omega)} + CM\|B\|_{L^2(\Omega)} \int_0^T \|z(s)\|_{L^2(\Omega)} ds.$$

Using the Gronwal inequality, we get

$$\|z_u(t)\|_{L^2(\Omega)} \leq C_1 \exp\left(CM\|B\|_{L^2(\Omega)} T\right). \tag{11}$$

with $C_1 = C\|z_0\|_{L^2(\Omega)}$.

On the other hand, the set $\{J(u)|u \in U\}$ is non-empty and is bounded from below by 0.

Let $(u_k)_{k \in \mathbb{N}}$ be a minimizing sequence in U such that $\lim_{k \to \infty} J(u_k) = \inf_{h \in U} J(h)$.

Then $(J(u_k))_{k \in \mathbb{N}}$ is bounded. Since $\|u_k\|_{L^2(0,T;L^2(\Omega))} \leq \frac{2}{\beta} J(u_k)$ thus, $(u_k)_{k \in \mathbb{N}}$ is bounded.

Thus, there exists a subsequence still denoted $(u_k)_{k \in \mathbb{N}}$ that weakly converges to a limit $u^* \in L^2(0, T; L^2(\Omega))$.

Since U is closed and convex, $u^* \in U$.

Let z_{u_k}, z_{u^*} be the corresponding solutions of system (2) to u_k and u^*, we have

$$z_{u_k}(t) - z_{u^*}(t) = \int_0^T S(t - s)[u_k(s)Bz_{u^*}(s) - u^*(s)Bz_{u^*}(s)]ds, \tag{12}$$

$$= \int_0^T S(t - s)\left[(u_k - u^*)(s)Bz_{u^*}(s) - u_k(s)(Bz_{u^*} - Bz_{u_k})(s)\right]ds, \tag{13}$$

This implies,

$$|z_{u_k} - z_{u^*}| \leq \left|\int_0^T S(t - s)(u_k - u^*)(s)Bz_{u^*}(s)ds\right| e^{\int_0^t \|S(t-s)\|\|u_k\|\|B\|ds} \tag{14}$$

Using the boudness of semigroup we get

$$|z_{u_k} - z_{u^*}| \leq C\left|\int_0^T S(t - s)(u_k - u^*)(s)Bz_{u^*}(s)ds\right|. \tag{15}$$

By theorem 3.9. in [4] the weak convergence $u_k \rightharpoonup u^*$ gives $u_k B z_{u^*}(.) \rightharpoonup$ $u^* B z_{u^*}(.)$ weakly in $L^2(0,T;L^2(\Omega))$.

Since $(S(t))_{t \geq 0}$ is compact, we have

$$\lim_{n \to \infty} \sup_{0 \leq t \leq T} |S(t-s)(u_k(s) - u^*(s))Bz(s)ds| = 0 \tag{16}$$

It follows that $z_{u_k} \to z^*$ strongly in $L^2(0,T;L^2(\Omega))$.

Since for $\alpha \in]0,1[$, D_x^α is continuous from $H_0^1(\Omega) \to L^2(\Omega)$, then

$$\lim_{k \to \infty} \int_0^T \|D_x^\alpha z_{u_k}(t) - z_d\|_{L^2(\Omega)} dt = \int_0^T \|D_x^\alpha z_{u^*}(t) - z_d\|_{L^2(\Omega)} dt.$$

and as J is lower, semi-continuous with respect to weak convergence, we have

$$J(u^*) \leq \lim_{k \to \infty} \inf J(u_k), \tag{17}$$

leading to $J(u^*) = \inf_{u \in U} J(u_k)..$

Remark 1.

If we consider the system (2) with a source term $f \in L^2(0,T;L^2(\Omega))$

$$\frac{dz}{dt} = Az + u(t)Bz + f \text{ on } Q \tag{18}$$

the same well-posedness and regularity results as hold, but the constant C_1 in Eq. (7) takes the form as follows:

$$C_1 = C\left(\|z_0\|_{L^2(\Omega)} + \|f\|_{L^2(0,T;L^2(\Omega))}\right).$$

3. Characterization

We now derive necessary conditions that an optimal control must satisfy. To derive these necessary conditions, we differentiate the cost functional. The differentiation result provides a characterization of the unique optimal control in terms of the optimality system.

In the next, we consider problem (2) and we discuses special cases of the set of admissible controls U.

Proposition 1.4.

Let consider the adjoint system given by:

$$\begin{cases} \dfrac{\partial p}{\partial t}(x,t) = -A^* p(x,t) + B^*(up)(x,t) + (D_x^\alpha)^* z_d(x) - (D_x^\alpha)^* D_x^\alpha z(x,t) & Q \\ p(x,t) = 0 & \Gamma, \\ p(x,T) = 0 & \Omega. \end{cases} \tag{19}$$

where z_u solution of system (2) and $(D_x^\alpha)^*$ is the adjoint operator of D_x^α.

Then the Frechet derivative of J at $u \in U$ is given by:

$$J'(u)(t) = p(t)Bz_u(t) + \varepsilon u(t).\tag{20}$$

Proof:

The system (12) has a weak solution $p \in L^2(0,T; L^2(\Omega))$ see [8], that satisfies:

$$p(t) = \int_t^T S^*(T-s)\left[B^*(up)(s) + (D_x^\alpha)^* z_d - (D_x^\alpha)^* D_x^\alpha z_u(s)\right]ds,\tag{21}$$

where $(S^*(t))_{t \geq 0}$ denotes the C_0 semi-group of generator $-A^*$, and B^* the adjoint operator of B.

Let consider the following system:

$$\begin{cases} \dfrac{\partial y}{\partial t}(x,t) = Ay(x,t) + u(x,t)By(x,t) + h(x,t)Bz_u(x,t) & Q, \\ y(x,t) = 0 & \Gamma, \\ y(x,0) = 0 & \Omega, \end{cases}\tag{22}$$

Let show that the mapping $\Psi : u \to z$ from $U \to L^2(0,T; L^2(\Omega))$ is Frechet differentiable, and $y = \Psi'(u).h$ is solution of system (15).

The operator $L : h \to y$ from U to W is linear.

Using remark (1) we have

$$\|y\|_{L^2(0,T;L^2(\Omega))} \leq C\|hBz_u\|_{L^2(0,T;L^2(\Omega))} \leq C_3\|z_u\|_{L^2(0,T;L^2(\Omega))},$$

It follows that L is continuous.

Now to show that Ψ is Frechet differentiable, it suffices to prove that

$$\lim_{\|h\|_U \to 0} \frac{\|\Psi(u+h) - \Psi(u) - L(h)\|_{L^2(0,T;L^2(\Omega))}}{\|h\|_{L^2(0,T;L^2(\Omega))}} = 0.$$

Setting $z_h = \Theta(u+h)$, $\psi = z_h - z_u$. and $\Phi = \psi - y$, then ψ and Φ are solutions of the following systems

$$\begin{cases} \dfrac{\partial \psi}{\partial t}(x,t) = A\psi(x,t) + u(x,t)B\psi(x,t) + h(x,t)Bz_h(x,t) & Q, \\ \psi = 0 & \Gamma, \\ \psi(x,0) = 0 & \Omega, \end{cases}\tag{23}$$

and

$$\begin{cases} \dfrac{\partial \Phi}{\partial t}(x,t) = A\Phi(x,t)u(x,t)B\Phi(x,t) + h(x,t)B\psi(x,t) & Q, \\ \Phi(x,t) == 0 & \Gamma, \\ \Phi(x,0) = 0 & \Omega, \end{cases}\tag{24}$$

It follows that

$$\|\psi\|_{L^2(0,T;H^1(\Omega))} \leq C\|h\|_{L^2(0,T;L^2(\Omega))}.$$

and

$$\|\Phi\|_{L^2(0,T;H^1(\Omega))} \leq \|hB\psi\|_{L^2(\Gamma)} \leq C\|h\|_U \|\psi\|_{L^2(0,T;H^1(\Omega))}.$$

Then $\|\Phi\|_{L^2(0,T;H^1(\Omega))} \le C\|h\|_U^2$.

It means that

$$\|\Theta(u+h) - \Theta(u) - \Theta'(u).h\|_{L^2(0,T;H^1(\Omega))} \le C\|h\|_U^2.$$

We conclude that Θ is Fréchet differentiable.

Let consider $u, u+h \in U$, then

$$\frac{1}{2}\|D_x^\alpha z_u - z_d\|_{L^2(0,T;L^2(\Omega))}^2 - \frac{1}{2}\|D_x^\alpha z_{u+h} - z_d\|_{L^2(0,T;L^2(\Omega))}^2 \tag{25}$$

$$= \int_0^T <D_x^\alpha(z_{u+h}(t) - z_u(t)), D_x^\alpha(z_{u+h}(t) - z_u(t)) - 2z_d>_{L^2(\Omega)}dt \tag{26}$$

$$= \int_0^T <z_{u+h}(t) - z_u(t), (D_x^\alpha)^* D_x^\alpha(z_{u+h}(t) - z_u(t)) - 2(D_x^\alpha)^* z_d>_{L^2(\Omega)}dt \tag{27}$$

$$= \int_0^T <y_h(t), (D_x^\alpha)^* D_x^\alpha(z_{u+h}(t) - z_u(t)) - (D_x^\alpha)^* z_d>_{L^2(\Omega)}dt + o\|h\|, \tag{28}$$

and

$$\frac{\beta}{2}\left(\|u+h\|_{L^2(0,T;L^2(\Omega))}^2 - \|u\|_{L^2(0,T;L^2(\Omega))}^2\right) = \beta<u,h>_{L^2(0,T;L^2(\Omega))} + o\|h\|.$$

Then J is Fréchet differentiable, and its derivative is given by:

$$J'(u).h = \int_0^T <y_h(t), (D_x^\alpha)^* D_x^\alpha(z_{u+h}(t) - z_u(t)) - (D_x^\alpha)^* z_d>_{L^2(\Omega)}dt + \beta<u,$$
$$h>_{L^2(0,T;L^2(\Omega))} + o\|h\|$$

Using the system (15), we have

$$\left\langle y_h(t), (D_x^\alpha)^* D_x^\alpha(z_{u+h}(t) - z_u(t)) - (D_x^\alpha)^* z_d\right\rangle_{L^2(\Omega)} \tag{29}$$

$$= \left\langle \int_0^T S(t-s)(u(s)By(s) + h(s)Bz(s))ds, (D_x^\alpha)^* D_x^\alpha(z_{u+h}(t) - z_u(t)) - (D_x^\alpha)^* z_d\right\rangle_{L^2(\Omega)} \tag{30}$$

Using the Gronwall lemma, we get

$$\int_0^T \left\langle y_h(t)(D_x^\alpha)^* D_x^\alpha(z_{u+h}(t) - z_u(t)) - (D_x^\alpha)^* z_d\right\rangle_{L^2(\Omega)}$$

$$= \left\langle \int_0^T \int_s^T S(t-s)(u(s)By(s) + h(s)Bz(s))dtds, (D_x^\alpha)^* \right.$$

$$\left. \times D_x^\alpha(z_{u+h}(t) - z_u(t)) - (D_x^\alpha)^* z_d\right\rangle_{L^2(\Omega)}$$

$$= \left\langle \int_0^T (u(s)By(s) + h(s)Bz(s))dt, \int_s^T S^*(t-s)(D_x^\alpha)^* D_x^\alpha(z_{u+h}(t) - z_u(t)) \right.$$

$$\left. - (D_x^\alpha)^* z_d \, dt\right\rangle_{L^2(\Omega)}$$

A variational formulation of system (12) leads to:

$$\int_s^T S^*(t-s)\left[(D_x^\alpha)^* z_d - (D_x^\alpha)^* D_x^\alpha z_u(s)\right]ds = p(s) - \int_s^T S^*(T-s)B^*(up)(t)dt, \quad (31)$$

It means that

$$\int_0^T \langle y_h(t), (D_x^\alpha)^* D_x^\alpha(z_{u+h}(t) - z_u(t)) - (D_x^\alpha)^* z_d\rangle_{L^2(\Omega)}$$

$$= \int_0^T \left\langle (u(s)By(s) + h(s)Bz(s)), p(s) - \int_s^T S^*(t-s)B^*(up)(t)dt, \right\rangle_{L^2(\Omega)} ds \text{Nonumber}$$

Using the Gronwall lemma once more gives

$$\int_0^T \left\langle (u(s)By(s) + h(s)Bz(s)), \int_s^T S^*(t-s)B^*(up)(s)dt \right\rangle_{L^2(\Omega)}$$

$$= \int_0^T \int_0^t \langle (u(s)By(s) + h(s)Bz(s)), S^*(t-s)B^*(up)(s)\rangle_{L^2(\Omega)} dsdt$$

$$= \int_0^T \int_0^t \langle (S(t-s)(u(s)By(s) + h(s)Bz(s)), B^*(up)(s)\rangle_{L^2(\Omega)} dsdt$$

$$= \int_0^T \left\langle \int_0^t S(t-s)(u(s)By(s) + h(s)Bz(s))ds, B^*(up)(t) \right\rangle_{L^2(\Omega)} dt$$

$$= \int_0^T \langle y(t), B^*(up)(t)\rangle_{L^2(\Omega)} dt = \int_0^T \langle u(t)By(t), p(t)\rangle_{L^2(\Omega)} dt$$

Then inequality (3) becomes

$$\int_0^T \langle y_h(t), (D_x^\alpha)^* D_x^\alpha(z_{u+h}(t) - z_u(t)) - (D_x^\alpha)^* z_d\rangle_{L^2(\Omega)} = \int_0^T \langle h(t)Bz(t), p(t),\rangle_{L^2(\Omega)} dt$$

Then the Frechet derivative of J is given by:

$$J'(u).h = \int_0^T \langle h(t), Bz(t)p(t)\rangle_{L^2(\Omega)} + \beta\langle u(t), h(t)\rangle_{L^2(\Omega)} dt.$$

The following results characterize and give an expression of an optimal control solution of problem (2) in several cases of admissible controls sets.
Proposition 1.5.
An optimal control solution of problem (2) is given by

$$u^*(x,t) = \max\left(m, \min\left(-\frac{1}{\beta}Bz(x,t)p(x,t), M\right)\right) \quad (32)$$

Proof:
The Frechet differential of J is given by

$$J'(u).h = \int_0^T \langle h(t)Bz(t), p(t)\rangle_{L^2(\Omega)} + \beta\langle u(t), h(t)\rangle_{L^2(\Omega)} dt.$$

Since J achieves its minimum at u^*, we have

$$0 \leq \int_0^T \langle h(t)Bz(t), p(t)\rangle_{L^2(\Omega)} + \beta\langle u(t), h(t)\rangle_{L^2(\Omega)} dt.$$

Taking $h = \max\left(m, \ \min\left(-\frac{1}{\beta}Bz(x,t)p(x,t), M\right)\right) - u^*$, we show that $h\left(u^* + \frac{1}{\beta}Bzp\right)$ is negative and then

$$\left(\max\left(m, \ \min\left(-\frac{1}{\beta}Bz(x,t)p(x,t), M\right)\right) - u^*\right)\left(u^* + \frac{1}{\beta}Bzp\right) = 0.$$

If $M \leq -\frac{1}{\beta}Bzp$ we have $(M - u^*)\left(u^* + \frac{1}{\beta}Bzp\right) = 0$, thus $u^* = M$.
If $m \leq -\frac{1}{\beta}Bzp \leq M$ we have $\left(-\frac{1}{\beta}Bzp - u^*\right)\left(u^* + \frac{1}{\beta}Bzp\right) = 0$.
Therefore $u^* = -\frac{1}{\beta}Bzp$.
Now, if $m \geq -\frac{1}{\beta}Bzp$, we have $(m - u^*)\left(u^* + \frac{1}{\beta}Bzp\right) = 0$ and then $u^* = m$.
We conclude that,

$$u^*(x,t) = \max\left(m, \ \min\left(-\frac{1}{\beta}Bz(x,t)p(x,t), M\right)\right).$$

The next proposition shows a necessary optimality condition.
Proposition 1.6.
Let $u^* \in U$ be an optimal control, then:

$$\forall v \in U, \quad <J'(u), u^* - v>_{L^2(0,T;L^2(\Omega))} \geq 0.$$

Proof:
If $v = u$, we get the condition.
If v is different than u, and since U is convex we have

$$u^* + \lambda(v - u^*) \in U, \quad \text{for any } \lambda \in]0,1[$$

It follows

$$J(u^*) \leq J(u^*) + \lambda(v - u^*)$$

which gives

$$J(u^*) \leq J(u^*) + \lambda <J'(u^*), v - u^*>_{L^2(0,T;L^2(\Omega))} + o(\lambda(v - u^*))$$

Then,

$$<J'(u^*), v - u^*>_{L^2(0,T;L^2(\Omega))} \geq \frac{1}{\lambda}(\lambda(v - u^*)).$$

Since $o(\lambda(v - u^*)) = \|\lambda(v - u^*)\|\varphi(\lambda(v - u^*))$, with $\lim_{\|z\|\to 0}\varphi(z) = 0$. Then

$$\lim_{\lambda\to 0}\frac{1}{\lambda}o(\lambda(v - u^*)) = \lim_{\|z\|\to 0}\|\lambda(v - u^*)\|\varphi(\lambda(v - u^*)) = \| v - u^*\|\lim_{\lambda\to 0}\varphi(\lambda(v - u^*)) = 0.$$

we conclude that,

$$<J'(u), u^* - v>_{L^2(0,T;L^2(\Omega))} \geq \lim_{\lambda \to 0} \frac{1}{\lambda} o(\lambda(v - u^*)) = 0.$$

Corollary 1.

Let $g \in L^2(\Omega)$, such that $|g| \mathbb{N}eq 0$ and assuming that $U = L^2(0,T)$. Then an optimal control is given by

$$u^*(x,t) = v^*(t)g(x) \tag{33}$$

with $v^*(t) = -\frac{1}{\beta \|g\|_{L^2(\Omega)}} \int_\Omega Bz(x,t)p(x,t)$

Particularly, if $g(x) = 1_D(x)$, with $D \subset \Omega$ is the actuator location and 1_D is the characteristic function such that its measure $\mu(D)$ is non-zero, then an optimal control $v^*(t)$ is given by

$$v^*(t) = \max\left(m, \min\left(-\frac{1}{\beta\mu(D)}\int_\Omega Bz(x,t)p(x,t)dx, M\right)\right). \tag{34}$$

Proof:

Let $v \in L^2(0,T)$, such that $w(x,t) = v(t)g(x)$ it follows from (1.6) that $\langle J'(u^*), w \rangle_{L^2(0,T;L^2(\Omega))} = 0$ which gives

$$\int_0^T v(t) \int_\Omega g(x)J'(u^*)(x,t)dxdt = 0 \quad \forall v \in L^2(0,T)$$

Hence

$$\int_\Omega g(x)J'(u^*)(x,t)dx = 0 \quad \forall t \in]0, T[$$

Then $\langle J'(u^*)(t), g \rangle_{L^2(\Omega)} = 0$, it means

$$\langle Bz(t)p(t), g \rangle_{L^2(\Omega)} + \beta v^*(t)\langle g, g \rangle_{L^2(\Omega)}, \quad \forall t \in]0, T[$$

which leads to formula (25).

4. Algorithm and simulations

In this section, we give an example to illustrate the usefulness of our main results.

The optimality condition (25) shows that the optimal control u^* is a function of z and p which themselves are functions of u^*. Then the control cannot be directly computed. For this reason, we introduce the following algorithm.

- Step 1: Choose an initial control $u_0 \in U$ a threshold accuracy $\varepsilon > 0$, and initialize with $k = 0$;

- Step 2: Compute z_k, solution of (2) and p_k, solution of (12) relatively to v_k.

- Step 3: Compute

$$v_{k+1} = \max\left(m, \min\left(-\frac{1}{\beta\mu(D)}\int_{\Omega} Bz_k(x,t)p_k(x,t)dx, M\right)\right). \tag{35}$$

- Step 4: If $\|u_{k+1} - u_k\| > \varepsilon, k = k + 1$, go to step 2. Otherwise $u^* = u_k$

For simulations, we consider a bilinear system described by the equation:

$$\begin{cases} \dfrac{dz}{dt}(x,t) = \Delta z(x,t) + u(x,t)z(x,t) & \Omega \times]0,1[\\ z(0,t) = z(1,t) = 0 &]0,1[\\ z(x,0) = x(x-1) & \Omega; \end{cases} \tag{36}$$

We consider problem (2) with $\alpha = 0.2$ and $z_d(x) = 0.62x^3 + 1.7x^2 + 0.023$. Applying the above algorithm, we obtain the following figures (**Figures 1** and **2**):

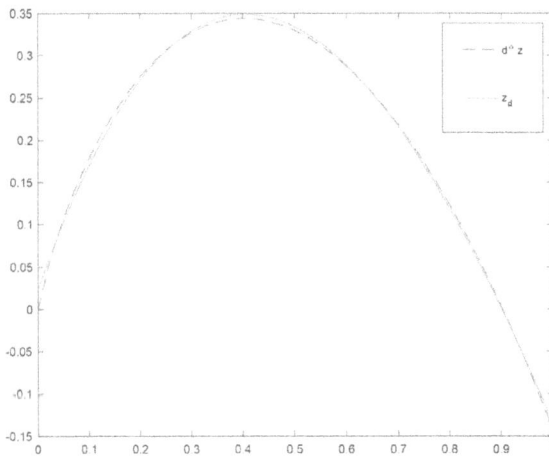

Figure 1.
Final state.

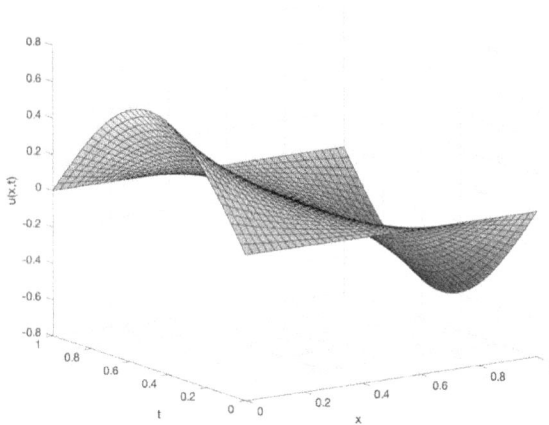

Figure 2.
The evolution of an optimal control.

The desired state is obtained with error $\|D_x^\alpha z(x,T) - z_d(x)\|_{L^2(\Omega)}^2 = 5.27.10^{-4}$ and a cost $J(u^*) = 2.31.10^{-3}$.

5. Conclusion

In this work, we discuss the question of fractional optimal control problem of parabolic bilinear systems with bounded controls, we obtain a distributed control solution, that minimizes a quadratic functional. This work gives an opening to other questions; this is the case of the fractional optimal control problem of hyperbolic systems. This will be the purpose of a future research paper.

Author details

Abella El Kabouss[*†] and El Hassan Zerrik[†]
MACS Team, Department of Mathematics, Moulay Ismail University, Meknes, Morocco

*Address all correspondence to: elkabouss.abella@gmail.com

† These authors contributed equally.

IntechOpen

References

[1] Hamayoun Stanikzai M, Elias S, Matsagar VA, Jain AK. Seismic response control of base-isolated buildings using tuned mass damper. Australian Journal of Structural Engineering. 2020;**21**(1): 310-321

[2] Khanna A, Kaur N. Vibration of non-homogeneous plate subject to thermal gradient. Journal of Low Frequency Noise, Vibration and Active Control. 2014;**33**(1):13-26

[3] Matin A, Elias S, Matsagar V. Distributed multiple tuned mass dampers for seismic response control in bridges. Proceedings of the Institution of Civil Engineers-Structures and Buildings. 2020;**173**(3):217-234

[4] Brezis H. Functional analysis, Sobolev spaces and partial differential equations. New York, Dordrecht, Heidelberg, London: Springer Science & Business Media; 2010

[5] Agrawal OP. Fractional optimal control of a distributed system using eigenfunctions. Journal of Computational and Nonlinear Dynamics. 2008;**3**:1-6

[6] Bahaa GM. Fractional optimal control problem for variational inequalities with control constraints. IMA Journal of Mathematical Control and Information. 2016;**6**:1-16

[7] Bahaa GM. Fractional optimal control problem for differential system with control constraints. Univerzitet u Nišu. 2016;**30**:2177-2189

[8] Bradley M, Lenhart S. Bilinear optimal control of a kirchhoff plate. Systems and Control Letters. 1994;**22**:27-38

[9] Bradley M, Lenhart S, Yong J. Bilinear optimal control of the velocity term in a kirchhoff plate equation. Journal of Mathematical Analysis and Applications. 1999;**238**:451-467

[10] Bradley ME, Lenhart S. Bilinear spatial control of the velocity term in a kirchhoff plate equation. Electronic Journal of Differential Equations. 2001;**2001**:1-15

[11] Lenhart S. Optimal control of a convective-diffusive fluid problem. Mathematical Models and Methods in Applied Sciences. 1995;**5**:225-237

[12] Addou A, Benbrik A. Existence and uniqueness of optimal control for a distributed-parameter bilinear system. Journal of dynamical and control systems. 2002;**8**:141-152

[13] Zerrik E, El Kabouss A. Regional optimal control of a class of bilinear systems. IMA Journal of Mathematical Control and Information. 2016;**34**: 1157-1175

[14] Liang M. Bilinear optimal control for a wave equation. Mathematical Models and Methods in Applied Sciences. 1999;**9**:45-68

[15] Lenhart S, Wilson D. Optimal control of a heat transfer problem with convective boundary condition. Journal of Optimization Theory and Applications. 1993;**79**:581-597

[16] Zerrik E, El Kabouss A. Regional optimal control of a bilinear wave equation. International Journal of Control. 2019;**92**:940-949

[17] Zerrik E, El Kabouss A. Bilinear boundary control problem of an output of parabolic systems. In: Recent Advances in Modeling, Analysis and Systems Control: Theoretical Aspects and Applications. Switzerland: Springer; 2020. pp. 93-203

[18] Zerrik E, El Kabouss A. Regional Optimal Control Problem of a Heat Equation with Bilinear Bounded Boundary Controls. In: Recent Advances in Intuitionistic Fuzzy Logic Systems and Mathematics. Cham: Springer; 2021. pp. 131-142.s

Chapter 3

Adaptive Sliding Mode Control Vibrations of Structures

Leyla Fali, Khaled Zizouni, Abdelkrim Saidi,
Ismail Khalil Bousserhane and Mohamed Djermane

Abstract

The sliding mode controller is one of the interesting classical nonlinear control-lers in structural vibration control. From its apparition, in the middle of the twen-tieth century, this controller was a subject of several studies and investigations. This controller was widely used in the control of various semi-active or active devices in the civil engineering area. Nevertheless, the sliding mode controller offered a low sensitivity to the uncertainties or the system condition variations despite the pres-ence of the Chattering defect. However, the adaptation law is one of the frequently used solutions to overcome this phenomenon offering the possibility to adapt the controller parameters according to the system variations and keeping the stability of the whole system assured. The chapter provides a sliding mode controller design reinforced by an adaptive law to control the desired state of an excited system. The performance of the adaptive controller is proved by numerical simulation results of a three-story excited structure.

Keywords: vibration control, sliding mode, adaptive law, earthquake excitation, Lyapunov stability

1. Introduction

The sliding mode controller is known as a powerful tool to control high-order complex nonlinear systems in presence of parametric uncertainty and external disturbances. The idea was initiated in the Soviet Union early in the 1930s [1, 2] after the Lyapunov stability theory apparition [3]. However, the peculiar evolution point started from the famous Emel'yanov and Barbashin works [4, 5]. Since then, the sliding mode control was a subject of several papers and works and been widely used in the various area as civil engineering [6], aircraft [7], robotic [8], energy and more other areas. After the contribution of the differential equations with discon-tinuous right-hand side theory established by Filippov in 1960 the sliding mode control received much more attention from researchers for wide dynamic system processes as time-varying, large-scale, infinite-dimensional or stochastic [9].

The sliding mode control decouples the dynamic motion of the whole controlled system into two components that do not depend on each other. In fact, this decom-position offered lower dimension and design simplicity to the system especially in feedback control conception. In addition to advantages presented by the sliding mode control as the insensitivity to parameter variations, complete rejection dis-turbances and depending on the sliding conditions the control can be combined easily to operational modes, approaches and controllers. Many interesting experimental and theoretical results are presented by combining the sliding mode control as adaptive sliding mode [10], fuzzy sliding mode [11], artificial neural network sliding mode [12] and decentralized sliding mode [13].

2. The concept of the sliding mode

The sliding mode aspect may appear in dynamical systems where the motion is presented using ordinary differential equations with discontinuous right-hand sides. Thus, the concept uses a discontinuous control signal to reform the system motions without depending on the system dynamic but the sliding parameters. This approach reduces the order of the original system equation which simplifies the mathematical modeling of the dynamic system motions. Therefore, the control output switched in high frequency between two values $\pm K$ and be subjected to discontinuities on the sliding surface in the state plane of the system to track the desired system state [14].

The linearization possibility of any mechanical system is linked to the presence of friction in the system. The force-velocity behavior depended essentially on the friction type as dry friction or fluid friction. In such a problem, the critical zone is which presented the maximum displacements. Nevertheless, in this zone the velocity value is in the neighborhood of zero with an opposite sign to the friction force. Consider the mechanical problem presented in **Figure 1** consisting of a Coulomb friction mass-spring system.

The motion equation is presented as

$$m\ddot{x} + c\dot{x} + kx = 0 \tag{1}$$

Where $\ddot{x}$, $\dot{x}$ and x are acceleration, velocity and displacement, m, c and k are the mass, friction coefficient and the spring stiffness.

In this case, the system work depends little on velocity and even if slowly moved the mass, a finite work is done in a displacement. So, even for the small velocity, the friction force existed and was defined with a finite value. Thereby, near to the zero velocity, the friction force switched to the finite limit in the two sides (positive or negative) [15].

$$c\dot{x} = \begin{cases} c_0\dot{x} & \text{with} \quad \dot{x} > 0 \\ -c_0\dot{x} & \text{with} \quad \dot{x} < 0 \end{cases} \tag{2}$$

c_0 is positive constant.

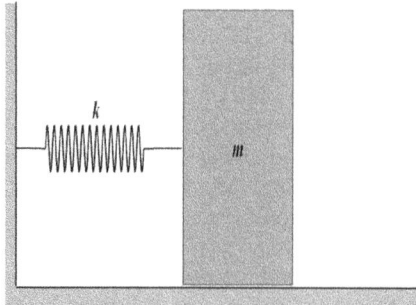

Figure 1.
Coulomb friction system.

Thus, around the state plan origin, the discontinuity is presented and the solution is unknown on the right-hand side of the differential motion equation. This situation is a frequent case in various control systems needing the differential equations with discontinuous right-hand sides theory [9].

The principle of the sliding mode approach consists of forcing the system to reach a fictive surface called the sliding surface to get the equilibrium state and keeping it switching around this surface thereafter. Hence, the first step is called the reaching phase and the second is called the sliding phase. Therefore, the trajectory in the state plane is assured by three distinct modes. The first mode is the convergence mode during which the variable to be adjusted gets the sliding surface from any initial point in the phase plane. This mode depended on the equivalent control law performance. The second mode is the smoothing mode in which the variable state reached the sliding surface and tends towards the origin of the state plane. The dynamic in this mode is characterized by the best choice of the sliding surface. The third one is the permanent regime mode characterizing the system response around the state plane origin depended on control law robustness. So, two steps to be followed are the determination of the fictive sliding surface on which the objectives of the controls are achieved then calculating the control law which ensures the state trajectory surface achievement and maintains it on this surface until reaching the state equilibrium [16].

3. Fundamental theory of the sliding mode control

The sliding mode control (SMC) is a two steps design controller in which the system motion is composed of two phases. The former step is the design of a fictive sliding surface to which the system motion must reach and hold the desired performances on it. However, the latter step is the design of the control law which drives the system motion to the sliding surface and maintains it on thereafter until reaching the equilibrium point. The sliding mode controller is a quake reacting controller. Whereas, the most advantage of this control that is insensitive to uncertainties or disturbances present in the system because the control design forces the system whatever to attain the surface prescriptions.

Let consider the general presentation of a nonlinear dynamic system as

$$\dot{x} = f(x,t) + g(x,t)u \qquad (3)$$

where x is the state vector, u is the desired control, $f(x)$ and $g(x)$ are nonzero smooth uncertain functions.

The sliding surface is presented as

$$\sigma = G \cdot e \qquad (4)$$

Where G is the sliding surface matrix and e is the tracking error of the system state defined as

$$e = x - x_0 \qquad (5)$$

x_0 is the desired state response.

Furthermore, to push the dynamic motion to the sliding surface the following conditions must be satisfied

$$\begin{cases} \sigma(x) = 0 \\ \dot{\sigma} = \dfrac{\partial \sigma}{\partial x}\dot{x} = \dfrac{\partial \sigma}{\partial x}(f(x,t) + g(x,t)) = 0 \end{cases} \tag{6}$$

The solution of the equation is the equivalent control of the sliding mode given by

$$u_{eq}(x,t) = -(\sigma(x)g(x,t))^{-1}\sigma(x)f(x,t) \tag{7}$$

To assure the sliding mode existence (σ = 0) the second component of the control law have to satisfy the attractively condition verified by

$$\dot{\sigma}(x) \cdot \sigma(x) < 0 \tag{8}$$

Because of the discontinuity presented on the sliding surface when the system reaches it and the Cauchy-Lipschitz theorem of the ordinary differential equations cannot be used [17]. The solution describing the dynamic behavior in this zone is using several approaches as the Filippov approach [18] or the Utkin approach [19] or more others as [20]. The above condition results

$$u = \begin{cases} u^+(x,t) & if & \sigma > 0 \\ u^-(x,t) & if & \sigma < 0 \end{cases} \tag{9}$$

From Eqs. (8) and (9) the sliding surface function and its derivative are of reverse sign and the second part of the sliding mode controller is given by

$$u_s = K \cdot sgn(\sigma) \tag{10}$$

Where K is the switching control gain and $sgn(\sigma)$ is the sliding surface signum function expressed by

$$sgn(\sigma) = \begin{cases} 1 & if & \sigma > 0 \\ 0 & if & \sigma = 0 \\ -1 & if & \sigma < 0 \end{cases} \tag{11}$$

Finally the sliding mode control law is presented as

$$u_{SMC} = u_{eq} + u_s \tag{12}$$

4. Sliding mode adaptation

Despite the claimed robustness properties of the sliding mode control, chattering is the harmful phenomenon affecting the control stability. This phenomenon is caused by the finite frequency oscillation of the switching part of the sliding controller. The presence of chattering in sliding mode control degrades the system accuracy and leads to the stability breaking and pushing the control to the divergence. Therefore, several researches and investigations focused on the chattering suppress methods and analysis. However, most of the chattering suppress methods consist of a continuous approximation of the discontinuous in the sliding surface neighborhood. The saturation is one of the main methods used in the chattering elimination in which a thin boundary layer around the surface is introduced defined as

$$sat(\sigma) = \begin{cases} \dfrac{\sigma}{\phi} & if & |\sigma| < \phi \\ sgn\,(\sigma) & & otherwise \end{cases} \qquad (13)$$

The boundary layer attributes the solution continuity and pushes the system to converge to this bound. The size of this layer depended on the system precision and the control accuracy. Another way to overcome chattering consists of the switching gain adaptation depending on the performance control maintain.

The adaptation is the ability of the system to adjust itself to its environment. Being processed, the adaptive system compensates the performance by changing its parameters depending on the plant environment evolution. Although, all the automatic adjustments in real-time approaches are considered as adaptive approaches with which the desired performance is maintained despite the system changes in time. Nevertheless, the adaptation of the sliding mode controller attenuates both the discontinuity and the chattering problem effect by adjusting the adapted gain depending on the plant environment. Thus, the adaptive approach is proposed to the switching part of the sliding mode controller of Eq. (10) and the equivalent part of Eq. (7) is maintained.

The adaptation of the controller in sliding mode consists of modifying in real-time the limit of the sliding boundary layer. While, a large band allows the system to regain the sliding surface easily but it destabilizes the controller by the length jump of its excessive gain. On the other hand, a small band causes a difficulty for the system to regain the sliding surface but it stabilizes the controller by the short jump of its low gain. Therefore, the proposed adaptive part is written as

$$u_{as} = \hat{K} \cdot sgn\,(\sigma) \qquad (14)$$

Consequently, the Eq. (12) becomes

$$u_{ASMC} = u_{eq} + u_{as} \qquad (15)$$

Where u_{ASMC} is the adapted sliding mode control law, u_{as} is the adapted switching part of the adaptive control, $\hat{K}$ is the new gain of the adaptive control proposed as

$$\hat{K} = \overline{K} - (\overline{K} - K) \cdot e^{-\alpha|\sigma|} \qquad (16)$$

$$\overline{K} = \lambda \cdot K \qquad (17)$$

Where $\overline{K}$ the amplified control gain, K is the original control gain of the Eq. (10), α is the convergence constant and λ is the amplification constant.

5. Lyapunov stability analysis

The convergence of the proposed adaptation law is evaluated and proved using the mathematical stability analysis of Lyapunov. Wherefore, The Lyapunov candidate function is chosen as [16].

$$V = \frac{1}{2}\sigma^2 + \frac{1}{2}\hat{e}^2 \qquad (18)$$

Where $\hat{e}$ is the adaptation error given by

$$\hat{e} = \hat{K} - K_d \tag{19}$$

Where K_d is the maximum value of $\hat{K}$ given by Eq. (16) with $K_d > d$ and d is the localized uncertainty related to the switching motion.

The first derivation of the candidate function can be presented as

$$\dot{V} = \sigma\dot{\sigma} + \hat{e}\,\dot{\hat{e}} \tag{20}$$

From Eqs. (17), (18) and (20) the above equation becomes

$$\dot{V} = \sigma(e\dot{e}) + \left(\hat{K} - K_d\right)\sigma \cdot sgn\,(\sigma) \tag{21}$$

$$e = K - \hat{K} = (\lambda - 1)K \cdot e^{-\alpha|\sigma|} \tag{22}$$

$$\dot{e} = -\alpha(\lambda - 1)K \cdot e^{-\alpha|\sigma|} \tag{23}$$

Thus, the Eq. (21) becomes

$$\dot{V} = \sigma\left(-\alpha(\lambda - 1)^2 K^2 \cdot e^{-2\alpha|\sigma|}\right) + \lambda\left(\hat{K} - K_d\right)\sigma \cdot sgn\,(\sigma) \tag{24}$$

$$\dot{V} = -\alpha(\lambda - 1)^2 K^2 \cdot e^{-2\alpha|\sigma|} \cdot \sigma - K_d \cdot |\sigma| \tag{25}$$

With $\lambda \geq 1$ and $0 < \alpha < 1$ the condition stability $V \cdot \dot{V}$ is verified.

6. Numerical examples

6.1 Single degree of freedom system

In order to evaluate the proposed adaptive sliding mode controller, we considered a single degree of freedom system composed of a spring-mass-damper system presented in **Figure** 2. Moreover, the system can move in the horizontal direction only and the influence of the adaptive nonlinear control responses of the vibrating system is evaluated [21]. In this example, the response of the system to a constant reference with an initial condition is presented (i.e. $x(0) = 0.11$). The parameters arbitrarily chosen of the spring-mass-damper system are: $m = 2$, $k = 1$ and $c = 0.5$.

The equilibrium force of the time-varying system is given by

$$f_I(t) + f_D(t) + f_S(t) = f_E(t) \tag{26}$$

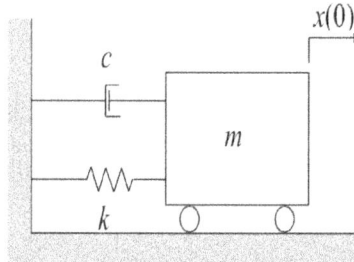

Figure 2.
Single degree of freedom example.

Where $f_I(t)$, $f_D(t)$, $f_S(t)$ and $f_E(t)$ are respectively the force of inertial, damping, spring and the external applied force given by

$$f_I(t) = m \cdot \ddot{x}(t) \tag{27}$$

$$f_D(t) = c \cdot \dot{x}(t) \tag{28}$$

$$f_S(t) = k \cdot x(t) \tag{29}$$

$$f_E(t) = k \cdot x(0) + f_c \tag{30}$$

Introducing Eqs. (27)-(30) in Eq. (26) yields

$$m \cdot \ddot{x}(t) + c \cdot \dot{x}(t) + k \cdot x(t) = k \cdot x(0) + f_c \tag{31}$$

Where m, c and k are the mass, damping and stiffness, $\ddot{x}$, $\dot{x}$ and x are acceleration, velocity and displacement, $x(0)$ and f_c are the initial displacement and the control required force.

Using Eqs. (12) and (15) to calculate respectively the classical sliding mode control and the adaptive sliding mode control forces. Thus, the two cases are simulated and compared to evaluate the robustness of the two controllers. Thereby, the numerical simulation result presented in **Figure 3** clearly shown the chattering reduction in the state plan response. This phenomenon is visibly reduced by using adaptive sliding mode control compared to the use of the classical sliding mode controller. Also, the adapted switching is clearly shown in the state plan presentation where the boundary layer thickness varied by the adaptation law depending on

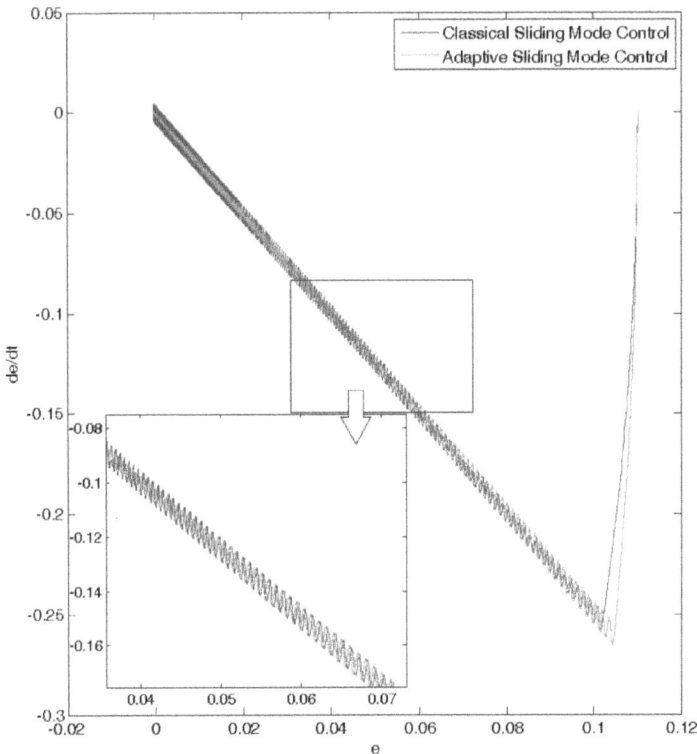

Figure 3.
Numerical simulation result of the mass-spring example.

the required performance of the plant. Besides, the **Figure 4** presented the compared numerical simulation results of the displacement responses function of time under the classical sliding mode control and the adaptive sliding mode control. Nonetheless, the displacement response of the system proves the performance of the adaptive nonlinear controller compared to the classical controller.

6.2 Multiple degree of freedom system

The previous example proved the efficiency of the used adaptive law to reinforce the control robustness. Otherwise, the applied load is a simple periodic load and the system is a simple system in which the switching output can be clearly shown in the state plan. In the present example, the three degrees of freedom system is considered under base excitation using earthquake records. This system is presented in **Figure 5** and the dynamic motion is governed by the following equation [22].

$$[M]\{\ddot{x}\} + [C]\{\dot{x}\} + [K]\{x\} = [M]\Lambda \ddot{x}_g + \{f_c\} \tag{32}$$

Where $[M]$, $[C]$ and $[K]$ are the mass, damping and stiffness matrices of the system, $\{\ddot{x}\}$, $\{\dot{x}\}$ and $\{x\}$ are the acceleration, velocity and displacement vectors, Λ, $\ddot{x}_g$ and $\{f_c\}$ are the load position vector, the one-dimensional earthquake vector and the control force vector. The example matrices of mass, damping and stiffness defined as [6, 23] are given respectively by

$$[M] = \begin{bmatrix} m_1 & 0 & 0 \\ 0 & m_2 & 0 \\ 0 & 0 & m_3 \end{bmatrix} = \begin{bmatrix} 98.3 & 0 & 0 \\ 0 & 98.3 & 0 \\ 0 & 0 & 98.3 \end{bmatrix} (kg) \tag{33}$$

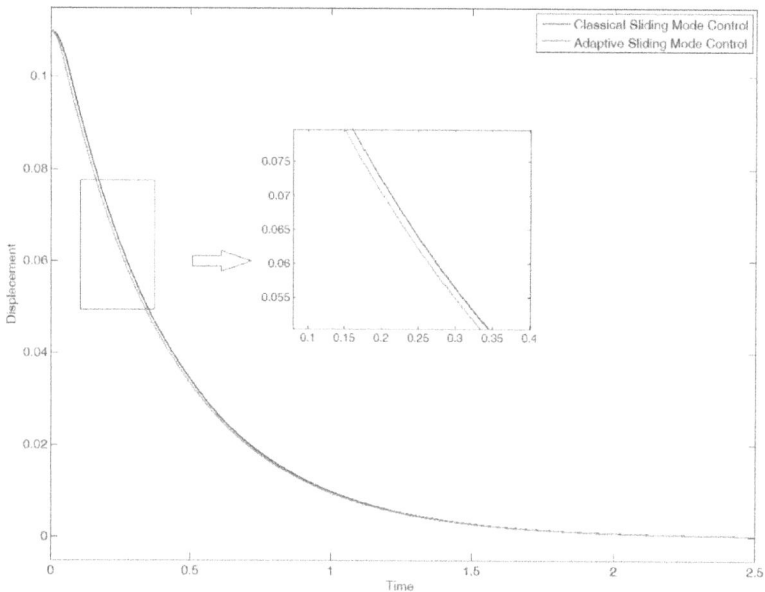

Figure 4.
Displacement responses of the mass-spring example.

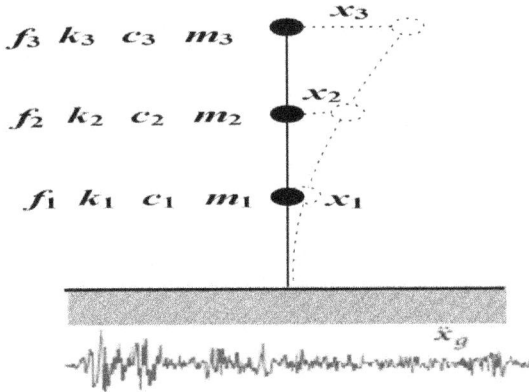

Figure 5.
Multiple degree of freedom example.

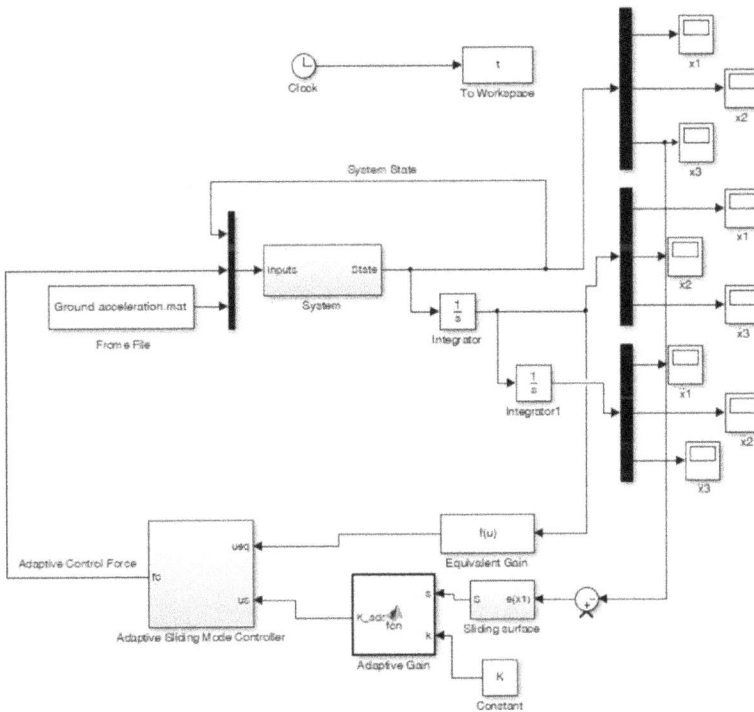

Figure 6.
Block diagram of the control system example.

$$[C] = \begin{bmatrix} c_1 + c_2 & -c_2 & 0 \\ -c_2 & c_2 + c_3 & -c_3 \\ 0 & -c_3 & c_3 \end{bmatrix} = \begin{bmatrix} 175 & -50 & 0 \\ -50 & 100 & -50 \\ 0 & -50 & 50 \end{bmatrix} (N \cdot s/m) \qquad (34)$$

$$[K] = \begin{bmatrix} k_1 + k_2 & -k_2 & 0 \\ -k_2 & k_2 + k_3 & -k_3 \\ 0 & -k_3 & k_3 \end{bmatrix} = 10^5 \begin{bmatrix} 12 & -6.84 & 0 \\ -6.84 & 13.7 & -6.84 \\ 0 & -6.84 & 6.84 \end{bmatrix} (N/m) \qquad (35)$$

Accordingly, the control example is achieved as presented in **Figure 6** and the required control force is calculated in a closed-loop forcing the system to reach the equilibrium state.

The system is excited using the scaled time of the Tōhoku 2011 earthquake record illustrated in **Figure 7**.

Although, to prove the effectiveness of the proposed adaptive sliding mode controller to suppress the structural vibrations of the excited system the numerical simulation results of the controlled and the uncontrolled system are compared. The displacement responses of the first mass of the structure of the two cases controlled and uncontrolled are shown in **Figure 8**. However, the second and the third mass displacement responses of the compared cases of numerical simulation are presented respectively in **Figures 9** and **10**. In addition, the inter-mass drift responses of the three masses are depicted in **Figure 11** in which the numerical simulation results of the uncontrolled system are compared to those of the adaptive controlled system. The adaptation of the switching gain value function of time under the 2011 Tōhoku earthquake excitation is presented in **Figure 12**.

From **Figures 8–10** the displacement responses are clearly reduced under the earthquake excitation. The inter-mass drift responses depicted in **Figure 11** show a remarkable reduction between the two cases controlled and uncontrolled systems. Moreover, the responses of the switching gain of the proposed adaptive law illustrated in **Figure 12** show the dependence on the excitation. For example, in

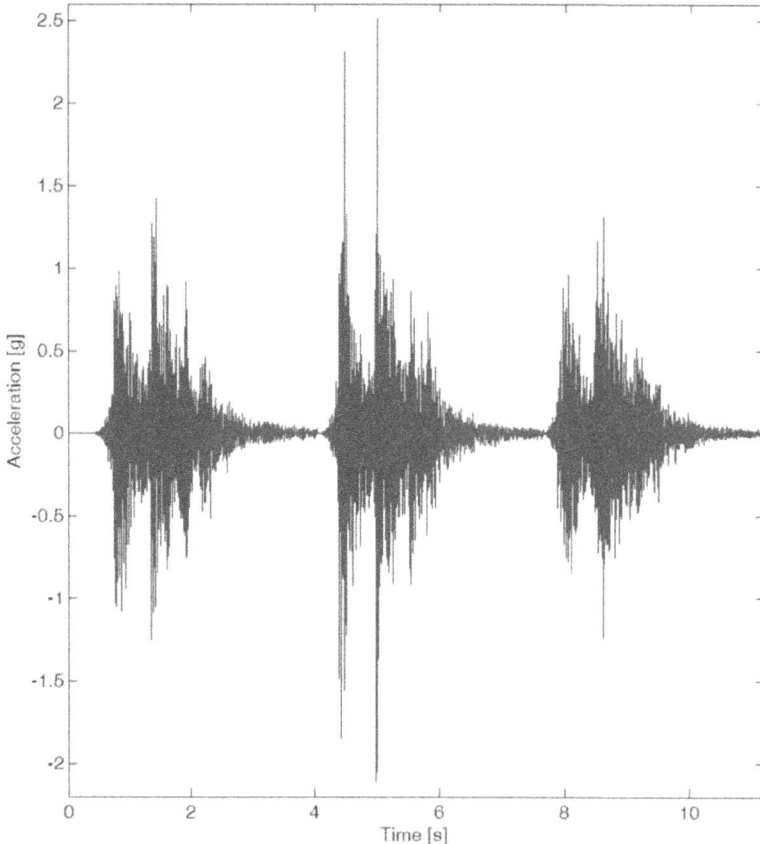

Figure 7.
The time scaled record of the 2011 Tōhoku earthquake.

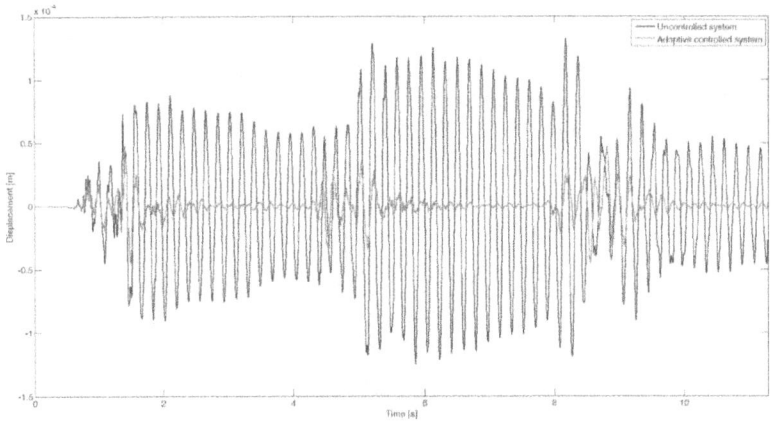

Figure 8.
The time displacement responses of the first mass under the 2011 Tōhoku earthquake.

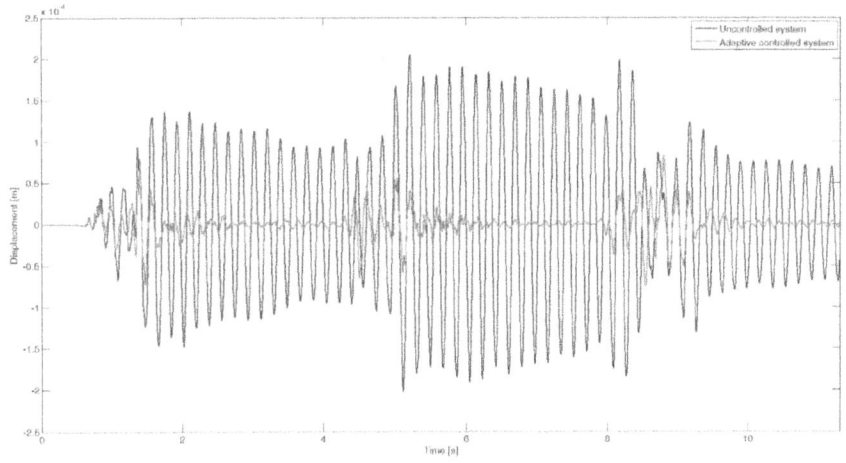

Figure 9.
The time displacement responses of the second mass under the 2011 Tōhoku earthquake.

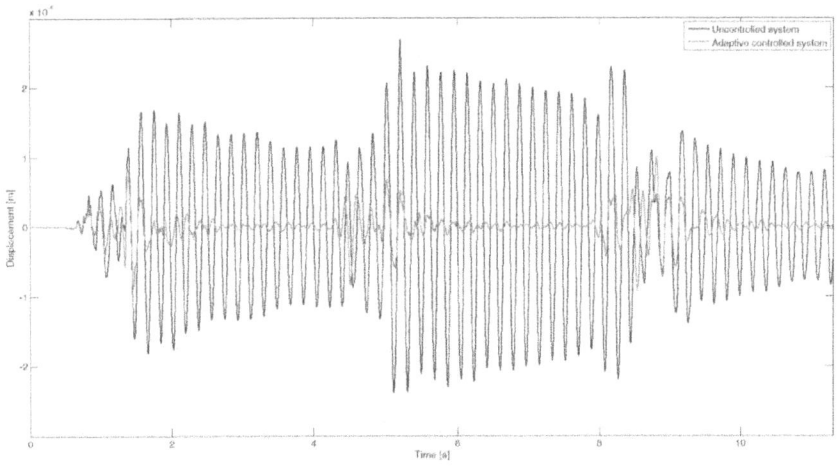

Figure 10.
The time displacement responses of the third mass under the 2011 Tōhoku earthquake.

Figure 12 between 4 and 5s where the peak seismic acceleration is located the law augmented the gain to the maximum value to track the system state better.

Over and above, the proposed adaptive control is evaluated by the result values of the system control application in the above-mentioned example. Some indexes are calculated and regrouped in **Table 1** to prove the robustness of the adaptive

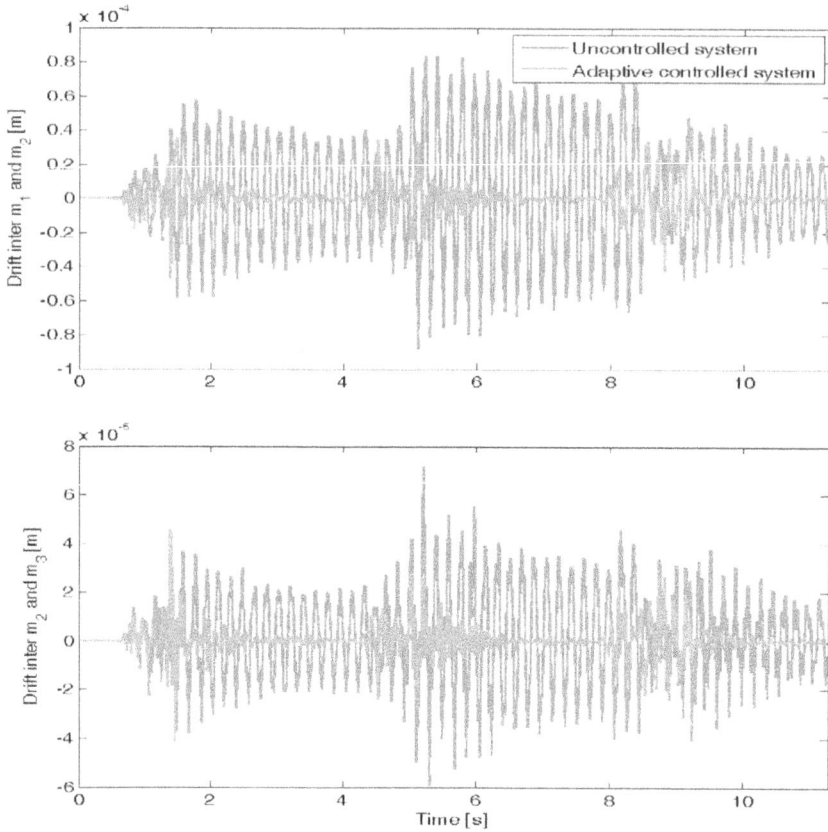

Figure 11.
The inter-mass drift responses under the 2011 Tōhoku earthquake.

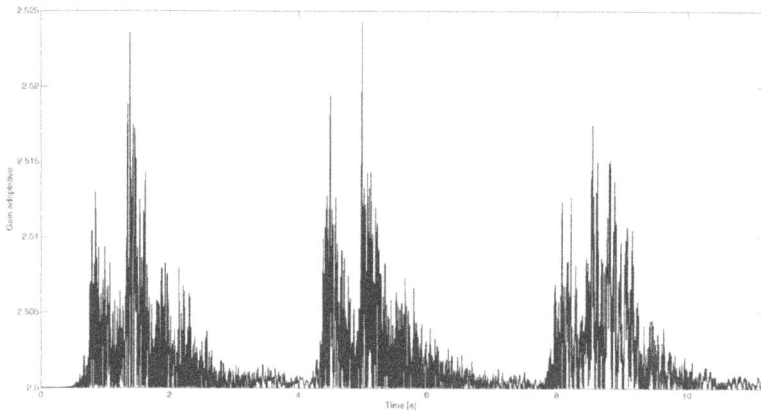

Figure 12.
The time adaptive gain variations under the 2011 Tōhoku earthquake.

Index	Formula	Mass number	Value (%)
Peak		1	65.33
Displacement	$\lvert x_i^{max}\rvert - max\,\lvert x_i\rvert/\lvert x_i^{max}\rvert$	2	58.78
Reduction		3	62.98
Peak		1	00.92
Acceleration	$\lvert \ddot{x}_i^{max}\rvert - max\,\lvert \ddot{x}_i\rvert/\lvert \ddot{x}_i^{max}\rvert$	2	02.79
Reduction		3	01.14
Peak drift	$\lvert d_i^{max}\rvert - max\,\lvert d_i\rvert/\lvert d_i^{max}\rvert$	1–2	56.63
Reduction		2–3	54.17

Table 1.
Calculated control indexes of the adaptive sliding mode control under the 2011 Tōhoku earthquake excitation.

control to attenuate the excited system vibrations. The peak displacement reduction and the peak acceleration reduction of each mass are calculated and inserted in **Table 1**. Thereby, the peak inter-mass drift reduction is also needful to evaluate the proposed adaptive controller performance.

Where the index i designed the mass number, x_i^{max} is the maximum uncontrolled mass displacement, x_i the controlled mass displacement, $\ddot{x}_i^{max}$ is the maximum uncontrolled mass acceleration, $\ddot{x}_i$ the controlled mass acceleration, d_i^{max} is the maximum uncontrolled inter-mass drift and d_i the controlled inter mass drift.

7. Conclusions

The proposed adaptive sliding mode controller robustness had been proved in the present chapter using two numerical examples. Although, the single degree of freedom example excited by a simple periodic load shown clearly the Chattering reduction as a result of the adaptive law effect. The numerical simulation results of the state plan presentation shown the switching gain adaptation value depending on the excitation effect. Moreover, the second example is a three degree of freedom system excited using an earthquake excitation to assure the presence of multiple frequencies and amplitudes. As expected, the numerical simulation results of the example prove the efficiency of the proposed adaptive controller. The peak mass displacement ratio attained 65.33%, consequently, the peak inter-mass drift is reduced by 56.63%. The peak acceleration is sparsely reduced because the adaptive control is designed to track the displacement only. In this stage, the nonlinear adaptive controller proves its effectiveness and performance in addition to the insensitivity to uncertainties or disturbances and system stability.

Nomenclature

C	Damping matrix
c	Damping or friction coefficient
c_0	Positive constant
d	Localized Uncertainty
d_i	Controlled inter-mass drift
d_i^{max}	Maximum uncontrolled inter-mass drift
e	Tracking error
$\hat{e}$	Adaptive error
f	Function
f_E	External force
f_D	Damping force

f_I	Inertial force
f_S	Spring force
f_c	Control force
G	Sliding surface matrix
g	Function
i	Mass number index
K	Switching gain, Stiffness matrix
$\hat{K}$	Adaptive switching gain
$\overline{K}$	Amplified switching gain
K_d	Maximum value of the adaptive switching gain
k	Stiffness
M	Mass matrix
m	Mass
u	The system output
u_{ASMC}	Adaptive sliding mode controller output
u_{SMC}	Sliding mode controller output
u_{eq}	Equivalent output
u_s	Switching output
u_{as}	Adaptive switching output
V	Lyapunov candidate function
x	Displacement
x_i	Controlled mass displacement
x_0	Desired response
$\dot{x}$	Velocity
$\ddot{x}$	Acceleration
$\ddot{x}_i$	Controlled mass acceleration
$\ddot{x}_i^{max}$	Maximum uncontrolled mass acceleration
σ	Sliding surface
ϕ	Boundary layer thickness
∂	Partial derivative
λ	Constant amplification
α	Convergence constant

Abbreviations

ASMC	Adaptive Sliding Mode Control
SMC	Sliding Mode Control
sat	Saturation function
sgn	Signum function

Author details

Leyla Fali[1*†], Khaled Zizouni[2†], Abdelkrim Saidi[2†], Ismail Khalil Bousserhane[2†]
and Mohamed Djermane[1†]

1 FIMAS Laboratory, TAHRI Mohamed University, Bechar, Algeria

2 ArchiPEL Laboratory, TAHRI Mohamed University, Bechar, Algeria

*Address all correspondence to: fali.leyla@univ-bechar.dz

† These authors contributed equally.

IntechOpen

References

[1] Kulebakin V. On theory of vibration controller for electric machines. Theoretical and Experimental Electronics. 1932;4. (In Russian)

[2] Nikolski G. On automatic stability of a ship on a given course. Proceedings of the Central Communication Laboratory. 1934;1:34–75. (In Russian)

[3] Lyapunov AM. The general problem of the stability of motion. Kharkov Mathematical Society. 1892. (In Russian)

[4] Emel'yanov SV. The way of obtaining complicated laws of control using only an error signal and its first derivative. Avtom. i Telemekh. 1957;18 (10):873–885. (In Russian)

[5] Barbashin E, Tabueva V, Eidinov R. On stability of variable control system under breaking conditions of sliding mode. Avtom. i Telemekh. 1963;24(7): 882–890. (In Russian)

[6] Fali L, Sadek Y, Djermiane M, Zizouni K. Nonlinear Vibrations Control of Structure under Dynamic loads. The 4th Student Symposium on Application Engineering of Mechanics SSAEM'4; 27-28 December 2018; Bechar, Algeria.

[7] Zhao J, Jiang B, Shi P, He Z. Fault tolerant control for damaged aircraft based on sliding mode control scheme. International Journal of Innovative Computing, Information and Control, 2014; 10(1); 293–302.

[8] Islam S, Liu XP. Robust sliding mode control for robot manipulators. IEEE Transactions on Industrial Electronics. 2010; 58(6): 2444–2453. DOI: 10.1109/ TIE.2010.2062472.

[9] Filippov AF. Differential equations with discontinuous Right-Hand Side. Matematicheskii Sbornik. 1960; 51(1): 99–128. (In Russian)

[10] Saidi A, Zizouni K, Kadri B, Fali L, Bousserhane IK. Adaptive sliding mode control for semiactive structural vibration control. Studies in Informatics and Control. 2019; 28(4):371–380. DOI: 10.24846/v28i4y201901.

[11] Bousserhane IK, Hazzab A, Laoufi A, Rahli M. Adaptive Fuzzy Sliding mode Controller for Induction Motor Control. 2nd International Conference on Information Communication Technologies; 24-28 April 2006; Damascus. p. 163–168. DOI: 10.1109/ICTTA.2006.1684363.

[12] Bouchiba B, Bousserhane IK, Fellah MK, Hazzab A. Artificial neural network sliding mode control for multi-machine web winding system. Revue Roumaine des Sciences Techniques - Serie Électrotechnique et Énergétique. 2017;62(1):109–113.

[13] Yan XG, Edwards C, Spurgeon SK. Decentralized robust sliding mode control for a class of nonlinear interconnected systems by static output feedback. Automatica. 2004; 40(4):613–620. DOI: 10.1016/j. automatica.2003.10.025.

[14] Sabanovic A, Fridman LM, Spurgeon S, Spurgeon SK. Variable structure systems: from principles to implementation. The Institution of Engineering and Technology Publication; 2004. 429 p. DOI: 978-0-86341-350-6

[15] Utkin VI. Sliding modes in control and optimization. Springer Science Business Media; 1992. 299 p. DOI: lO.1007/978-3-642-84379-2

[16] Fali L. Semi-active vibration attenuation of civil engineering structures using an adaptive nonlinear law. Bechar: TAHRI Mohamed University; 2020.

[17] Amann H. Ordinary Differential Equations: An Introduction to Nonlinear Analysis. Walter De Gruyter Inc; 1990. 458 p. DOI: 10.1515/9783110853698

[18] Filippov AF. Differential Equations with Discontinuous Right-hand Sides. Kluwer Academic Publishers; 1988. 304 p. DOI: 10.1007/978-94-015-7793-9

[19] Utkin IV. Sliding mode in control optimization. Springer-Verlag, Berlin; 1992. 286 p. DOI: 0-387-53516-0

[20] Levaggi L, Villa S. On the regularization of sliding modes. SIAM Journal on Optimization, 2007;18(3): 878–894. DOI: 10.1137/060657157

[21] Elias S, Matsagar V. Research developments in vibration control of structures using passive tuned mass dampers. Annual Reviews in Control. 2017; 44: 129–156. DOI: 10.1016/j.arcontrol.2017.09.015

[22] Stanikzai MH, Elias S, Matsagar VA, Jain AK. Seismic response control of base-isolated buildings using tuned mass damper. Australian Journal of Structural Engineering. 2020; 21(1): 310–321. DOI: 10.1080/13287982.2019.1635307

[23] Fali L, Djermane M, Zizouni K, Sadek Y. Adaptive sliding mode vibrations control for civil engineering earthquake excited structures. International Journal of Dynamics and Control. 2019; 7(3):955–965. DOI: 10.1007/s40435-019-00559-0

Development of a Low-Cost Vibration Damper Dynamometer for Suspension Damper Testing

Yucheng Liu and Ge He

Abstract

On performance vehicles, suspension dampers are used to reduce the vibration produced by variations in the driving surface, while simultaneously controlling the rate of load transfer between tires during lateral and longitudinal acceleration. To measure the characteristics of suspension dampers, a damper dynamometer is typically used to compress and elongate the dampers at a known speed, and then measure the force output. However, a commercial damper dynamometer is usually expensive and not always suitable for the dampers specifically designed for a customized vehicle. In this chapter, a cheap, customized, and effective damper dynamometer is constructed through computer-aided design, finite element analysis, and manufacture to measure the properties of suspension dampers used in a racecar. It was demonstrated through data analysis that this designed damper dynamometer can produce usable measurement data for a far lower cost than other methods.

Keywords: suspension damper, damper dynamometer, computer-aided design, finite element analysis

1. Introduction

Dampers are being successfully and widely used to reduce vibrations in most applications, such as civil engineering structures and automotive components. In civil engineering [1–4], for example, adding fluid viscous dampers to buildings can help protect buildings, bridges, and other structures in a variety of scenarios including seismic events, strong winds, and pedestrian energy. For automotive engineering, proper suspension damping reduces the vibration produced by variations in the driving surface, while simultaneously controlling the rate of load transfer between tires during lateral and longitudinal acceleration. Because these damping modes occur at different speeds of compression and rebound (elongation), the best racing dampers offer damping rate adjustments at both high and low speeds.

For at least the past 4 years, our team has used Öhlins TTX25 MkII dampers on its racecars, seen in **Figure 1**. These dampers retail at $650 each, but they include the high- and low-speed damping rate adjustments necessary for optimal damper performance. For an independent suspension vehicle, this comes to a total cost of $2600, offering the team a significant financial incentive to reuse them between design cycles. Luckily, the manufacturer offers detailed documentation on how

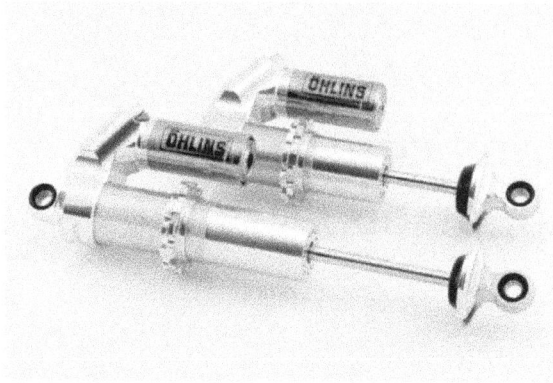

Figure 1.
Öhlins TTX25 MkII dampers.

to perform proper maintenance, so the team performs full damper rebuilds when reusing the dampers on a new car.

One drawback of rebuilding dampers is the inability to easily tell whether their performance will remain unchanged after the rebuild. This can lead to different damping rates on each corner of the car, resulting in less than ideal performance of the car's suspension. To prevent this, it is important to measure the characteristics of each damper after rebuilding it. However, unlike springs, the damping characteristics cannot be determined from simple static measurements and sophisticated devices,

Figure 2.
Intercomp 3HP shock dyno 1.0–55 in/s ($8895.00).

such as damper dynamometers are required to correctly measure them [5–7]. A damper dynamometer is a specialized machine that compresses and elongates the dampers at a known speed and then measures the force output. Market offerings for damper dynamometers are well outside the team's price range, with all viable options costing thousands of dollars, such as the Intercomp model shown in **Figure 2**.

These devices facilitate easy and straightforward measurement and data processing, but the trade-off in price is too high for the team to justify for such a specialty tool. An alternative option offered online is mailing the rebuilt dampers to a company specializing in damping rate measurement. Priced at around $600 total, this route represents a significant decrease in cost, but still a relatively high yearly expense for the team. In addition, this would only enable the team to measure their dampers in a single setting, eliminating much of the team's ability to accurately compare the effects of different damping rates on the car. In light of the limitations and costs associated with the commercial damper measurement options available to the team, it was determined that the best course of action would be to design and manufacture a custom damper dynamometer catered to the specific needs of the team. Three primary requirements were established, in order of importance. While the custom damper dynamometer is temporarily used for characterizing the vehicle dampers, it is expected that the same design method can also be adopted for designing a custom dynamometer measuring damping rates of civil engineering structures. The design method described in this paper can also help increase the efficiency in designing dampers that are used for vibration control.

2. Design requirements

The most important requirement was that the dynamometer must produce usable measurement data. This was the primary purpose of the project and so takes priority over all other goals. The desired output of the dynamometer is a relationship between the compression/rebound rate and the resistive force output by the damper. Data should be measured at a rate of approximately 100 Hz for each sensor, comparable to the typical rate of data measurement the team uses when logging track data. The range of necessary compression and rebound speeds varies by damper; however, the FSAE team is primarily concerned with speeds up to 10 in/s. This corresponds with the maximum tested speed in the available force-velocity data for the Öhlins TTX25 MkII. The measurement device should maintain linearity up to at least 250 lbf of damper resistive force, which is the maximum force output achieved in the available force-velocity data for the dampers.

The second requirement was that the design would utilize wherever possible components that the team already owned. This was to reduce the cost of the project as much as possible, important for keeping it a viable financial alternative to purchasing a dynamometer or sending off the team's dampers for measurement.

The third requirement was that all parts of the project were able to be completed within a single semester. As a single-semester-directed individual study, it was imperative that the project was approached in such a way that it would be completed before the deadline. This requirement was changed out of necessity due to the COVID-19 pandemic and resultant changes to the accessibility of university purchasing and machining resources. Though the timeframe of completion shifted, there was still only around a semester of available time to work on this project. Due to this, the design was required to be simple enough that almost everything could be made in-house, reducing the lead time that would result from having to order components.

3. Design

The design of the dynamometer, as seen in **Figure 3**, was obtained following an existing system engineering design process [8–14] and a combined experimental-computational approach [11, 12]. A crank-slider mechanism imparts a forced displacement to one end of the damper, while the other end is mounted to a canti-levered bar. A welded frame constructed from low-carbon steel angle stock holds the crank-slider mechanism together. In **Figure 3a**, some sections of the frame have been made transparent for ease of viewing other components, and the model does not include fastening hardware. **Figure 3b** shows the fabricated and setup dyna-mometer. Parallel alongside the damper is a linear potentiometer used on the team's racecars for the exact purpose of measuring damper displacement. Attached at the base of the cantilevered bar are two strain gauges to measure the strain in the bar, and indirectly, the resistive force of the damper. This is similar to the design of the Intercomp dynamometer, with several key changes to reduce the price.

For the crank mechanism, a section of a retired crankshaft from one of the team's old engines was utilized. The CBR600 engine it came from has a cylinder stroke of 42.5 mm, approximately 75% of the usable stroke of the damper, provid-ing the necessary leeway for setup adjustment. This crankshaft was modified to fit inside one of the milling machine's R8 collets. A steel rod was tapered and threaded to match an existing threaded hole in the crankshaft, in order to ensure collinearity during welding. Also utilized was the connecting rod, along with its big end bearing inserts and a section of the wrist pin. Since these parts already have tight tolerances on their interfacing surfaces, they offered a perfect opportunity to eliminate free play in the system while also cutting the cost and machining time.

The slider portion of the crank-slider mechanism consists of a smooth-surfaced rod with clevises on each end, constrained to a single degree of freedom with a custom aluminum bushing block. Careful surface preparation and lubrication allowed for the use of a bushing rather than a more expensive linear ball bearing while preventing mechanism binding, which would result in extra stresses in the frame and torque applied to the mill.

The welded frame was subjected to finite element analysis in SolidWorks to determine its adequacy for the maximum expected forces. The forces applied in the finite element model included the 250 lbf maximum damper output force, reacted by the bolt holes of the bearing which supports the crank. Conservatively, it was assumed that the entirety of the force was reacted by the frame, when in actuality, the mill will resist a portion of the force. In addition, the transverse component of the force in the connecting rod was calculated at its most extreme angle. This force (58 lbf), and its associated moment about the Z-axis (143 lbf-in), was applied to the

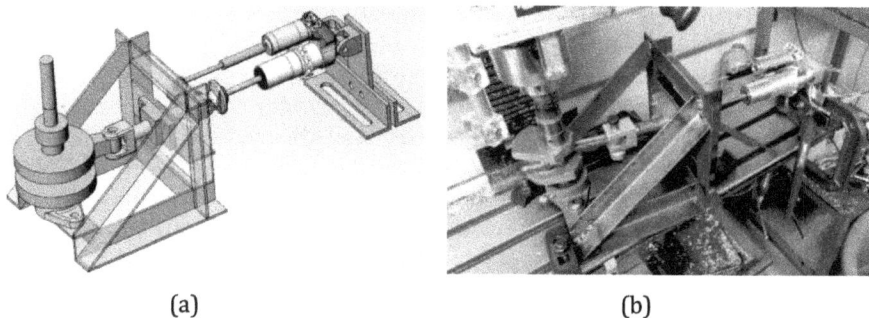

(a)   (b)

Figure 3.
(a) Damper dynamometer CAD model, (b) completed dynamometer setup.

Figure 4.
Finite element simulation of dynamometer frame.

aluminum bushing block that houses the slide rod. The finite element simulation result for this combined load case is shown in **Figure 4**.

As shown in **Figure 4**, the maximum stress experienced by the frame is 14.7 ksi. This corresponds to a factor of safety of approximately 3.7. At this point, further iteration could have reduced the weight of the frame. However, since weight was not a primary concern, it was decided that maintaining the thicker frame angles would result in easier welding operations. This will also allow the frame to potentially be used for testing larger dampers without modification.

The size of the cantilevered strain bar was chosen based on the expected output force of the dynamometer. For a set of available materials, the material thickness and width, as well as the magnitude and location of the applied force from the damper, were used to calculate the bending stress at the point where the strain bar was supported by its base. The chosen bar's thickness and width allow for a bending stress that is just below the yield stress for the material. At an applied force of 250 lbf and the designed cantilever moment arm of 2.375″, a bending moment of 625 lbf-in is generated. The selected bar is composed of 1018 steel (yield strength = 53.7 ksi), with a rectangular cross-section of 2″ width by 3/16″ thickness. At the supported point where the bending stress is highest, this corresponds to an applied bending stress of approximately 50.7 ksi, or approximately 94% of the material's yield strength. This allows the bar to fully react the maximum expected force without yielding while maximizing the detectable elastic strain in the beam at lower force outputs. For higher damper force output applications in the future, a different cantilevered bar may be needed.

Used to measure the strain in the bar from the resistive force in the damper are two foil resistance strain gauges, nominally 350 ohm. Near the supported end of the strain bar, as close as possible to the point of maximum strain, the surface was prepared using progressively finer sandpaper. Strain gauges were attached to either face of the strain bar using a cyanoacrylate adhesive. The strain gauges (STRG1 and STRG2) were wired in a Wheatstone half-bridge configuration, according to the circuit diagram shown in **Figure 5**.

Originally, the bridge was completed using 350 ohm resistors, but these were swapped for 47 kohm resistors to limit the excessive noise seen in that half of the bridge, and to allow for the use of an available rotary potentiometer (POT 1) to effectively balance the bridge. The result is a steadier measurement of bridge imbalance, and the ability to center the measurement circuit output within the

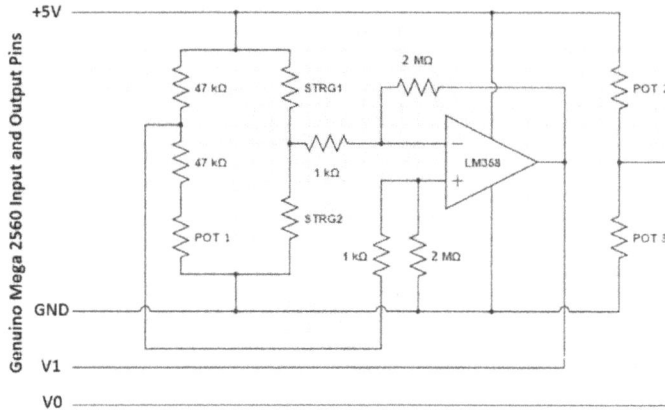

Figure 5.
Diagram of measurement circuit.

measurable range of a Genuino Mega 2560 used for data capture. Using the Genuino as a 5 V supply voltage, the bridge imbalance is adjusted via the rotary potentiometer to account for imperfections in the manufacture and connection of the strain gauges. The voltage difference created by the bridge circuit under load is amplified using an LM358 op-amp chip in a differential amplifier configuration, also shown in **Figure 5**. The gain of the amplifier circuit is set at 2000 using a combination of 2 Mohm and 1 kohm resistors, so that the output is within the readable range of the Genuino and provides a large enough measurable range.

The linear potentiometer that measures the instantaneous length of the damper is represented on the far-right side of **Figure 5** as two variable resistors (POT 2 and POT 3), one of which increases resistance with increasing length and one of which decreases. The potentiometer uses a 5 V supply from the Genuino to output a maximum signal at full extension and a minimum signal at full compression.

Two input pins on the Genuino board (V0 and V1) are used to measure the op-amp output and the linear potentiometer output. The program loaded onto the board runs in a loop, conveying with each iteration the timestamp in milliseconds as well as a value between 0 and 1023 for each input pin. These values correspond to the voltage at each input pin, with the 0–5 V measurement range broken up evenly into 1024 subdivisions. A delay written into the program is adjusted to provide data points at a rate of 100 Hz, satisfying the data capture requirement outlined earlier.

These values are transmitted as comma-delimited serial data to the user's computer via USB, and the program RealTerm Serial/TCP Terminal is used to capture the data. After identifying the correct computer port for the incoming data transmission, RealTerm allows the dynamometer user to write all captured data to a text file, which is then parsed into Microsoft Excel for further analysis.

4. Data analysis

The data is parsed into Microsoft Excel as a 4-column dataset. The data includes iteration number since the Genuino program began executing, timestamp in milliseconds since the program began executing, instantaneous voltage reading at input pin V0, and instantaneous voltage reading at pin V1. It should be noted that the raw data does not contain the initialization of the Genuino program, and thus the data does not begin at an iteration and timestamp of zero. This is acceptable because the iteration number is only included to ensure that no steps have been skipped and the program is running properly.

The first step in analyzing the data is to establish a baseline for the strain gauge circuit output. Before turning on the milling machine to drive the dynamometer, several seconds of data are captured to establish an accurate baseline. In addition, the machine is turned off and allowed to rest for several more seconds before the capture program is terminated, in order to determine if the baseline changed during operation. This is possible if some components of the circuitry shift during the use of the dynamometer and warrant further attention and possibly recapture of the dataset.

The circuit used to measure the applied damper force was calibrated by mounting the cantilevered strain bar onto a vertical surface and applying known loads up to 200 lbs. at the location of the damper attachment, in intervals of 50 lbs. The linear potentiometer was calibrated by measuring the voltage output and comparing it to a measurement of displacement, in intervals of approximately 0.25 in. The results of these calibrations are shown in **Figures 6** and 7.

Fitting a trend line to the calibration data shows that a high degree of linearity is maintained over the measured range. Because of the difficulty associated with accurately applying large known loads during calibration, it is necessary to assume that the linearity will hold true up to a load of 250 lbf. The high degree of linearity

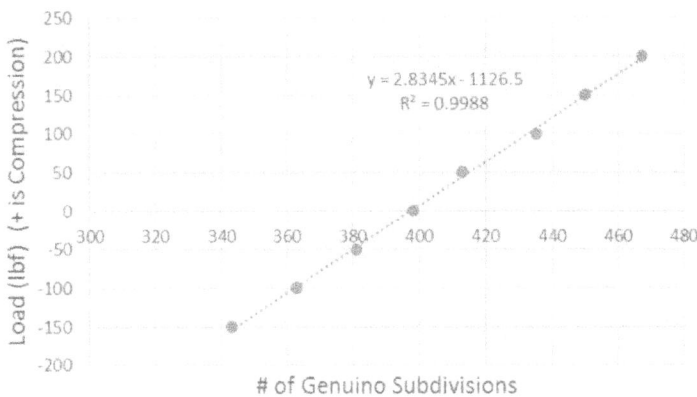

Figure 6.
Strain gauge calibration plot.

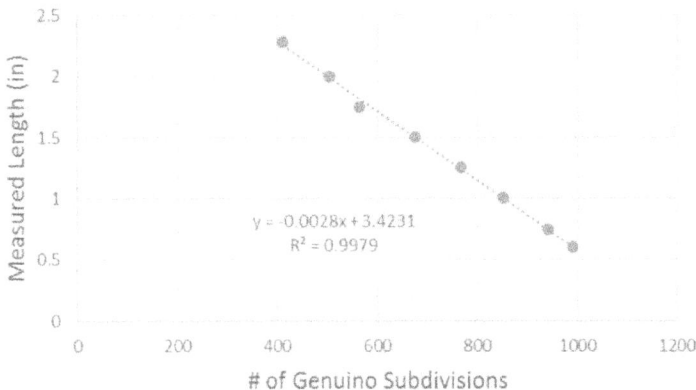

Figure 7.
Linear potentiometer calibration plot.

seen in the calibration data justifies this assumption. From this data, the user can interpolate or extrapolate the applied load or displacement for a given Genuino voltage measurement.

After the sensors have been calibrated, it is possible to accurately plot the displacement of the driven end of the damper as a function of time. It is assumed that the strain bar contributes a negligible amount of displacement to the nondriven end of the damper. However, because the distance between the damper connection points is measured directly by the linear potentiometer, even without this assumption the measurement should be accurate. By taking the derivative of the displacement with respect to time, the compression or rebound velocity of the damper is obtained for each timestamp.

Because of noise in the measurement circuit, larger datasets are required to produce a smooth force-velocity curve for the damper. To analyze these large quantities of data, the method that was chosen is to take the average force-displacement over a range of input velocities. For example, the force and velocity data at all points which indicate a compression velocity of between −0.25 in/s and 0.25 in/s is averaged to produce a single data point. The same is done for all points which indicate a compression velocity of between 0.25 in/s and 0.75 in/s, and so on, until the entire data set is accounted for. The same process is performed for rebound velocities.

Force-velocity graphs were generated for the range of adjustments listed in the available data from the manufacturer. The graphs sweep through a range of low-speed settings at the maximum high-speed setting, and a range of high-speed settings at the maximum low-speed setting. The naming convention of the graphs is chosen as LS-HS, where LS is low speed and HS is high speed. Low-speed settings are counted in clicks from fully closed, whereas high-speed settings are counted in revolutions from fully open. Thus, a graph labeled 0-3 shows the data for fully closed low-speed adjustments and 3 rotations on each high-speed adjustment. All graphs are included in **Figure 8**. It should be noted that a graph was not generated for the setting 0-4.3, because the miniature mill was unable to maintain the necessary velocity profile under high load. This is discussed further in the Issues and Future Improvements section of this paper. All settings were adjusted symmetrically to match the format of the published data accessible in [13].

The manufacturer-supplied curves published in [13] show the compression and rebound responses above and below the x-axis, respectively. After all measured data has been analyzed and plotted, it is possible to compare the measurements from the damper dynamometer to the manufacturer's published data. The graphs are first compared qualitatively, and it can be seen that there are certain observable similarities and differences between the plots. Like the manufacturer graphs, the measured force increases with higher settings, showing that the constructed dynamometer can clearly illustrate the difference between damper settings, and was able to measure the expected data trends. The measured data is visually different in the graphs of the 15-4.3, 25-4.3, and 0-0, in that there is a small velocity domain within the rebound response where the damping force decreases as the velocity increases. This is not seen in any of the manufacturer graphs and possibly suggests that the tests should be rerun. In addition, the quality of the different dynamometer systems can be seen in the graphed data. Because of noise in the measurement system of the constructed dynamometer, inconsistencies and discontinuities are common in the measured data, contrasted with the smoothly generated curves of the manufacturer data.

Quantitative analysis of the graphs allows for calculation and discussion of the error between the measured force output and that which is expected from the manufacturer data. For each run, the force output values are obtained from the measured

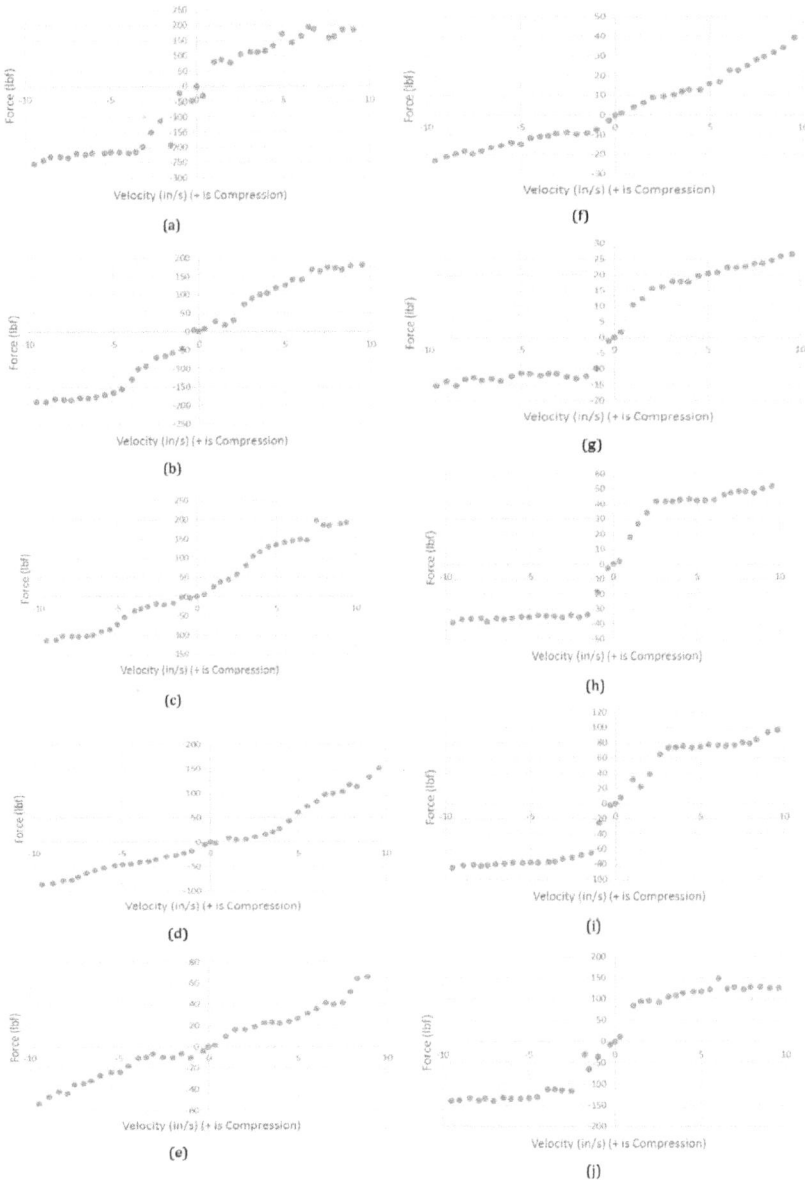

Figure 8.
Force-Velocity Curves from Measured Data. (a) 2-4.3, (b) 4-4.3, (c) 6-4.3, (d) 10-4.3, (e) 15-4.3, (f) 25-4.3, (g) 0-0, (h) 0-1, (i) 0-2, (j) 0-3.

data at velocities of 5 in/s and 10 in/s, and these are compared to graphically obtained values at the same velocities from the published data [13]. These values and the associated error calculations are shown below in **Table 1**.

From **Table 1**, it can first be noted that the error values are overwhelmingly positive. This clear trend suggests that either a difference exists between the team's damper and a new one from the manufacturer, or that the dynamometer was improperly calibrated. Further investigation is necessary to determine which of these factors is the cause of this discrepancy.

Dataset	Measured @ 5 in/s (lbf)	Measured @ 10 in/s (lbf)	Manuf. @ 5 in/s (lbf)	Manuf. @ 10 in/s (lbf)	Error @ 5 in/s	Error @ 10 in/s
0-0 C	20	26	18	24	0.11	0.08
0-0 R	11	15	13	16	−0.15	−0.06
0-1 C	42	52	44	49	−0.05	0.06
0-1 R	35	40	31	37	0.13	0.08
0-2 C	75	100	75	92	0.00	0.09
0-2 R	80	85	58	63	0.38	0.35
0-3 C	115	125	107	125	0.07	0.00
0-3 R	135	145	80	103	0.69	0.41
2-4.3 C	145	190	115	150	0.26	0.27
2-4.3 R	210	250	94	130	1.23	0.92
4-4.3 C	125	180	96	140	0.30	0.29
4-4.3 R	165	190	73	115	1.26	0.65
6-4.3 C	140	190	70	130	1.00	0.46
6-4.3 R	75	110	57	105	0.32	0.05
10-4.3 C	60	150	38	100	0.58	0.50
10-4.3 R	50	90	30	75	0.67	0.20
15-4.3 C	25	60	25	65	0.00	−0.08
15-4.3 R	23	55	15	42	0.53	0.31
25-4.3 C	17	40	13	35	0.31	0.14
25-4.3 R	15	23	10	23	0.50	0.00

Table 1.
Measured and manufacturer data and calculated error.

Cases averaged	Avg. error
Rebound	0.42
Compression	0.22
High-Speed sweep	0.14
Low-Speed sweep	0.44
Force <100 lbf	0.16
Force >100 lbf	0.51

Table 2.
Average error for comparison between case sets.

Comparison between the average percent errors for certain segments of data gives insight into the areas where the constructed dynamometer is most accurate. These average values are listed in **Table 2** for easy comparison.

From this table, it should be noted that the measured data conforms more accurately to the published manufacturer data, in compression, than in rebound and shows much higher variation at higher loads and during low-speed sweep test cases. Assuming the measured damper does have the performance characteristics of a new damper from the manufacturer, this variability between the datasets shows where the dynamometer is least accurate.

5. Issues and future improvements

The primary challenge associated with building this dynamometer at the lowest possible price point was integrating parts the team already owned rather than purchasing them, while at the same time creating a dynamometer that produces accurate measurements. Certain issues with the design could be solved with more work, and likely will be as the team continues to use the dynamometer, thus are discussed here.

Firstly, the measurement circuit should be improved to limit electronic noise in the measurements. As designed, all the circuitry between the sensors and the Genuino is constructed on a breadboard, which has the drawback of loose connections causing changes in the resistance of some circuit components during dynamometer operation. This was reduced where possible by using higher-resistance components, but a small movement in the connection of either of the nominal 350-ohm strain gauges can change their effective resistance by a significant percentage. This should be easily achievable by soldering resistors instead of using a breadboard. Limiting noise should provide better data for easier processing and would allow for more accurate determination of hysteresis present in the system, that may otherwise be overlooked.

The second primary issue with the dynamometer is that the miniature milling machine which provides the driving torque to the crank is unable to maintain a constant angular velocity under the load applied by the damper. In practice, this means the crank slows down when the connecting rod is at its most extreme angle to the damper and speeds up dramatically when the two are aligned. It is likely that this is the reason for the large percent error in higher-force tests. It is also hypothesized that this is one of the reasons for the large difference in error between rebound and compression cases. To address this issue, the team should consider obtaining permission to set up the damper dynamometer on a larger university milling machine. Not only would it likely be able to supply more torque and a more constant angular velocity, but it should also provide an overall stiffer framework for the dynamometer to operate within.

The present study can be converted into a course project for mechanical engineering students who take the vibrations and controls class to develop their hands-on experience and strengthen their understanding of the concepts of the dynamic behavior of vibration systems delivered in that class (**Table 3**) [14, 15].

Component	Price
Steel angles for frame	64.92
LM358 op-amp chips	6.99
Foil strain gauge sensors	13.99
Total	86.90

Table 3.
Price breakdown.

6. Conclusion

To improve the performance of the FSAE car's suspension, the goal of this project was to design and build a low-cost dynamometer capable of producing a force-velocity curve for the car's dampers. Through the use of primarily pre-owned components, this dynamometer was constructed for less than 1/100 the price of market alternatives. Primary differences include the use of the team's milling machine to drive the dynamometer crank rather than a dedicated motor, and a cantilevered strain bar with strain gauges, custom wiring, and a Genuino to measure the force, rather than a dedicated load cell and computer system. While these changes offer great financial savings, this project has shown that they are not without their drawbacks. The graphs require calibration and data processing to produce and do not exactly replicate the manufacturer published values for the damper characteristics. This project, however, was still successful. It provided a completed and usable damper dynamometer, which through further testing and refinement will be able to accurately determine the characteristics of the team's dampers for a far lower cost than other methods.

Conflict of interest

The authors declare no conflict of interest.

Author details

Yucheng Liu[1] and Ge He[2*]

1 South Dakota State University, Brookings, USA

2 University of Maryland, Baltimore, USA

*Address all correspondence to: gh663@shu.edu.cn

IntechOpen

References

[1] Stanikzai MH et al. Seismic response control of base-isolated buildings using tuned mass damper. Australian Journal of Structural Engineering. 2020;**21**(1):310-321

[2] Matin A, Elias S, Matsagar V. Distributed multiple tuned mass dampers for seismic response control in bridges. Proceedings of the Institution of Civil Engineers-Structures and Buildings. 2020;**173**(3):217-234

[3] Elias S. Effect of SSI on vibration control of structures with tuned vibration absorbers. Shock and Vibration. 2019;**2019**:12. Article ID: 7463031. DOI: 10.1155/2019/7463031

[4] Khanna A, Kaur N. Vibration of non-homogeneous plate subject to thermal gradient. Journal of Low Frequency Noise, Vibration and Active Control. 2014;**33**(1):13-26

[5] Liu Y-C. Modeling abstractions of vehicle suspension systems supporting the rigid body analysis. International Journal of Vehicle Structures & Systems. 2010;**2**(3-4):117-126

[6] Liu Y-C. Recent innovations in vehicle suspension systems. Recent Patents on Mechanical Engineering. 2008;**1**(3):206-210

[7] Liu Y-C. Design of instructional tools to facilitate understanding of fluid viscous dampers in a vibration and controls class and course assessment. In: 2020 ASEE Virtual Annual Conference. Washington, D.C.: American Society for Engineering Education; June 22-26. 2020

[8] Liu Y-C, Batte JA, Collins ZH, Bateman JN, Atkins J, Davis M, et al. Mechanical design, prototyping, and validation of a Martian robot mining system. SAE International Journal of Passenger Cars—Mechanical Systems. 2017;**10**(1):1289-1297

[9] Liu Y-C, Meghat V, Machen B. Design and prototyping of a debris clean and collection system for a cylinder block assembly conveying line following an engineering systems design approach. International Journal of Design Engineering. 2018;**8**(1):1-18

[10] Liu Y-C, Meghat V, Machen B. Design and prototyping of an *in situ* robot to clean a cylinder head conveying line following an engineering systems design approach. International Journal of Design Engineering. 2017;**7**(2):106-122

[11] Liu Y, Whitaker S, Hayes C, Logsdon J, McAfee L, Parker R. Establishment of an experimental-computational framework for promoting Project-based learning for vibrations and controls education. International Journal of Mechanical Engineering Education. 2022;**50**(1):158-175. DOI:10.1177/030641 9020950250

[12] Liu Y-C. Implementation of MATLAB/Simulink into a vibration and control course for mechanical engineering students. In: ASEE SE Section Annual Conference; Auburn University, Auburn, AL, USA, Washington, D.C.: American Society for Engineering Education; March 8-10. 2020

[13] Available from: http://www.ohlinsusa.com/files/TTX25%20 MkII%20Dyno%20lbs%20vs%20ips.pdf [Accessed: August 31, 2020]

[14] Liu Y-C, Baker F, He W-P, Lai W. Development, assessment and evaluation of laboratory experimentation for a mechanical vibrations and controls course. International Journal of Mechanical Engineering Education. 2019;**47**(4):315-337

[15] Liu Y-C, Baker F. Development of vibration and control systems through

student projects. In: ASEE SE Section
Annual Conference; North Carolina State
University, Raleigh, NS, USA,
Washington, D.C.: American Society for
Engineering Education; March 10-12. 2019

Chapter 5

A Ball-Type Passive Tuned Mass Vibration Absorber for Response Control of Structures under Harmonic Loading

Jiří Náprstek and Cyril Fischer

Abstract

Ball-type tuned mass absorbers are growing in popularity. They combine a multi-directional effect with compact dimensions, properties that make them attractive for use at slender structures prone to wind excitation. Their main drawback lies in limited adjustability of damping level to a prescribed value. Insufficient damping makes ball-type absorbers more prone than pendula to objectionable effects stemming from the non-linear character of the system. Thus, the structure and design of the damping device have to be made so that the autoparametric resonance states, occurrence of which depends on system parameters and properties of possible excitation, are avoided for safety reasons. This chapter summarises available 3D mathematical models of a ball-pendulum and introduces the non-linear approach based on the Appell–Gibbs function. Efficiency of the models is then illustrated for the case of kinematic and random excitation. Interaction of the absorber and the harmonically forced simple linear structure is numerically analysed. Finally, the chapter provides examples of typical patterns of the autoparametric response and outlines possibilities of applications in practical engineering.

Keywords: passive vibration damping, non-linear dynamics, autoparametric systems, semi-trivial solution, dynamic stability

1. Introduction

Design of contemporary structures is often distinguished by their slenderness, which is either functionally, economically or aesthetically motivated. Consequently, the structure lacks required stiffness and vibration absorbers are thus required to be incorporated into the structure design. The generally accepted family of passive tuned mass vibration absorbers is well established in the engineering literature; see the exhaustive review paper [1], which reflects the situation until 2017. The history of tuned vibration absorbers dates back to the beginning of the twentieth century [2]. However, the history of the analysis of non-linear effects connected with these devices is much younger. A tuned mass absorber represents a non-linear system, which, in connection with the supporting structure, has an autoparametric character (for general description of the topic, see [3]). As such, the system is prone to an autoparametric stability loss. This kind of problem was initially investigated in [4] and later elaborated in works by Tondl and others [5]. A new layer of complexity must be taken into account when both spatial components of the absorber are considered because the autoparametric interaction occurs between the two directions as well. This effect has been known for a long time (see the classical analysis of the chaotic behaviour of a spherical pendulum [6]), however, its relation to pendulum-based tuned mass absorbers was neglected until recently [3].

There are structures where installation of a classical pendulum-based absorber is not possible for spatial, aesthetic, or other reasons. Ball-type passive tuned mass absorbers, which are based on free movement of a heavy ball rolling in a spherical cavity, represent an alternative solution. They combine a multi-directional effect with compact dimensions and thus are convenient for use at towers, mast, footbridges, and other slender structures. For example, ball-type absorbers are increasingly popular in connection with wind turbines [7], namely, for offshore installations where the simultaneous influence of wind and wave loads makes the dynamic response of the turbines more complex. Moreover, such devices are almost maintenance-free; the importance of this property naturally increases with the number of installations [8]. Despite many advantages of ball-type absorbers, they have limited damping level adjustability. There are various techniques that implement additional damping into the absorber as a rubber coating or liquid introduced in the cavity. It also seems that the usage of several balls in one container may improve effectiveness of the absorber due to the impact effect and the rolling friction [7]. These modifications, however, entail significant maintenance costs with uncertain results. In any case, insufficient damping makes ball-type absorbers substantially more prone to objectionable effects stemming from the non-linear character of the system compared to pendula, namely, to an autoparametric-based energy transfer between individual components of the system. This effect can result in, for example, an increasing amplitude of the transverse motion of the absorber when only mild excitation takes place.

Mathematical modelling of the movement of a homogeneous sphere rolling on a perfectly rough surface has a long tradition in classical mechanics. The system is non-holonomic with linear constraints in the first derivatives with respect to time. The classical setting of several particular cases including rolling of a sphere on a spherical surface are considered in a classical monograph by Routh [9]. A similar Lagrangian approach was used in popular monographs [10] and is still used regularly [11]. As an alternative, the Appell–Gibbs approach, being based on an *energy acceleration function*, appears to be more effective in some cases, providing governing systems that are more transparent for further elaboration [12]. Abstract solutions using Lie group theoretical methods were derived recently [13]. This approach allows for generalisation of the cavity shape to non-symmetrical surfaces of the second order, however, it is less convenient for practical use.

The first papers dealing with theoretical, experimental and practical aspects of ball-type absorbers were published by Pirner [14]. His design procedure was based on a simplified planar approach. In a follow-up paper [15], the authors of this chapter modelled an absorber and a supporting structure as a non-linear planar structure. The detailed stability analysis of the complete system revealed the typical autoparametric behaviour exhibiting harmonic, chaotic or multi-valued response intervals, which can represent a dangerous state for the structure.

The spatial version of the absorber model is an autoparametric system where the longitudinal direction (parallel with the excitation movement) is supposed as the primary component and the lateral direction plays the role of a secondary component. If the system enters autoparametric resonance, vibration of the primary coordinate acts as parametric excitation of the secondary coordinate due to mutual non-linear relations. If the secondary component remains in rest and the primary one vibrates, the so-called *semi-trivial state* occurs. The interaction of an absorber with a structure adds a new degree of complexity to the system. The structure, being driven by an external forcing, adopts the role of the primary component; however, both coordinates of the absorber maintain their mutual relations. The particular states of such an autoparametric system are characterised by the existence of bifurcation points that delimit stable and unstable solution branches.

The authors of this chapter put forth significant effort in describing the autoparametric character of pendulum- and ball-type absorbers. For the ball-pendulum, the 2D approach based on the Lagrangian formalism [15] offered a possibility of a detailed analytic description of the reduced problem, where the stable and unstable response domains were clearly identified. The numerical evaluation of the 3D model derived using the Appell–Gibbs function according to [16] revealed important physical properties of the system and many particular trajectory types in forced and free movement cases [17, 18]. It was also found that the resonance properties of the 3D model are similar to those obtained analytically and experimentally for the spherical pendulum [19, 20]. These results support validity of the mathematical model and numerical analyses presented and used in this chapter.

The idea of the ball-pendulum serving as a vibration absorber offers wide possible generalisations. Apart from the aforementioned usage of multiple balls in a cavity or multiple stacked devices for damping multiple frequencies, usage of non-homogeneous spheres, nested spheres, hemishperes or semielliptic spheres would allow the absorber to be fine-tuned for a precisely limited non-linear damping effect or multidirectional damping. For example, an analysis of a Chaplygin ball on a spherical surface is presented in [21]; the bidirectional damping based on a rolling-pendulum is introduced in [22]. A significant disadvantage of nonhomogeneous systems is their weaker stability when compared to traditional symmetric devices. Alternatively, usage of cavities with a general shape may represent a more convenient alternative; see, for example, a case with a semielliptic cavity analysed by Legeza [23]. It is worth noting that survey [1] does not mention any paper regarding dynamic stability analysis of the vibration absorber-equipped structures, although this topic is mentioned as one requiring significant attention. It seems that the research work being conducted on this topic is currently aimed at non-linear dynamic absorbers with different non-linearities in damping, as those involving friction elements [24] or different kinds of non-linear springs [25].

The chapter is organised as follows. After this introduction, the chapter describes the governing differential system based on the Appell–Gibbs approach. Next, the chapter discusses the autoparametric behaviour of the absorber itself for harmonic and random kinematical excitation. Then, the chapter presents results from numerical simulation of the simplified structure equipped with the absorber. The last section of the chapter concludes.

2. Mathematical model

2.1 Appell–Gibbs function of the system

The mathematical model of a simplified structure equipped with a ball-type vibration absorber (see **Figure 1**) consists of two main components: the simplified dynamical model of the supporting structure and the absorber (i.e., the cavity with a ball of mass m). The absorber is connected to the structure at point A, which is

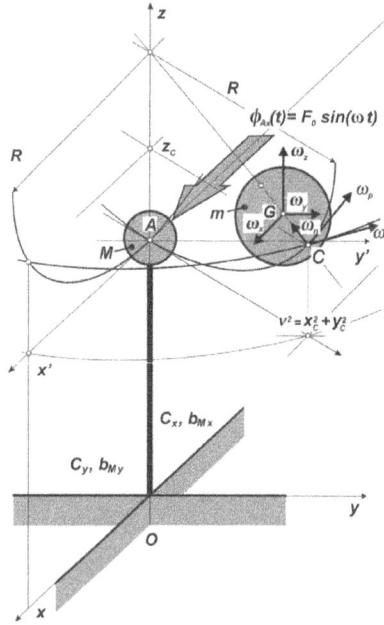

Figure 1.
The structure modelled as two SDOF subsystems together with a 3D ball vibration absorber.

supposed to move horizontally with respect to coordinates x, y; leaning of the structure and rotation around axis z are neglected.

Thus, the complete system includes eight degrees of freedom: two describing the movement of the top of the structure and six for the absorber, three of which are related by three non-holonomic constraints of the first order expressed in velocities. The detailed derivation of the model of the ball-type absorber was already published and thus we will only briefly summarise it here. For further information on the derivation, the reader is kindly referred to [26].

The system behaviour can be characterised by the Appell–Gibbs function where $\mathbf{u}_G = (u_{Gx}, u_{Gy}, u_{Gz})$ and $\mathbf{w} = (\omega_x, \omega_y, \omega_z)$ describes the translational and rotational movement of the ball in the cavity. Symbols $\mathbf{u}_A = (u_{Ax}, u_{Ay}, u_{Az})$ denote the displacement of the top of the structure, which is modelled as two SDOF linear damped oscillators representing movement of concentrated mass M independently in horizontal directions x, y:

$$S = \frac{1}{2}m\left(\ddot{u}_{Gx}^2 + \ddot{u}_{Gy}^2 + \ddot{u}_{Gz}^2\right) + \frac{1}{2}J\left(\dot{\omega}_x^2 + \dot{\omega}_y^2 + \dot{\omega}_z^2\right) + \frac{1}{2}M\left(\ddot{u}_{Ax}^2 + \ddot{u}_{Ay}^2\right), \quad (1)$$

$$M\ddot{u}_{Ax} + b_x\dot{u}_{Ax} + C_x u_{Ax} = \Phi_x,$$
$$M\ddot{u}_{Ay} + b_y\dot{u}_{Ay} + C_y u_{Ay} = \Phi_y, \quad (2)$$

$u_{Az} = 0$, where forces Φ_x and Φ_y comprise effect of loading and interaction with the absorber.

In function S, the first and second parts (m, J) represent dynamics of the ball moving within the cavity of the absorber, while the third term (M) refers to the structure together with the case of the absorber (without the ball), as shown in **Figure 1**.

The following notation was adopted in Eq. (1):

R, r—radius of the cavity or the ball, respectively, [m],

M, m—mass representing the structure including mass of the static part of the absorber, mass of the ball moving inside the cavity, respectively, [kg],

J—central inertia moment of the ball with respect to point G; parameter J allows to consider whatever type of spherical body with central symmetry, the mass of which is either concentrated in the centre ($J = 0$), uniformly distributed within the body ($J = 2/5mr^2$), or evenly dispersed mass over the outer envelope of the ball ($J = mr^2$), [kg m²],

ω—angular velocity vector of the ball with respect to its centre G, [rad s^{-1}],

A—moving origin related with the cavity in its bottom point,

u_A—displacement of the contact between the structure and the cavity,

u_G—displacement of the ball centre with respect to the moving origin A, [m],

C—contact point of the ball and cavity,

$u_C = (u_{Cx}, u_{Cy}, u_{Cz})$—displacement of the ball contact point with respect to the moving origin A, [m],

$x = x, y, z$—Cartesian coordinates with origin in the point O,

C_x, b_x, C_y, b_y—structure stiffness and linear viscous damping in x, y horizontal directions, [N m^{-1}, Ns m^{-1}].

2.2 Ball movement inside the cavity

From the supposition of a non-sliding contact between the ball and cavity, the velocities of the ball centre with respect to origin can be deduced providing the respective non-holonomic constraints of "perfect" rolling. Thus, the conditions for displacement vectors u_C and u_G can be written as:

$$\dot{u}_{Cx} = \omega_y(u_{Cz} - R) - \omega_z u_{Cy}, \qquad \dot{u}_{Gx} = \dot{u}_{Ax} + \rho(\omega_y(u_{Cz} - R) - \omega_z u_{Cy}),$$
$$\dot{u}_{Cy} = \omega_z u_{Cx} - \omega_x(u_{Cz} - R), \qquad \dot{u}_{Gy} = \dot{u}_{Ay} + \rho(\omega_z u_{Cx} - \omega_x(u_{Cz} - R)), \qquad (3)$$
$$\dot{u}_{Cz} = \omega_x u_{Cy} - \omega_y u_{Cx}, \qquad \dot{u}_{Gz} = \dot{u}_{Az} + \rho(\omega_x u_{Cy} - \omega_y u_{Cx}),$$
$$\text{where}: \ \rho = 1 - r/R \ \text{and} \ \dot{u}_{Az} = 0.$$

The ball centre acceleration $\ddot{u}_G = (\ddot{u}_{Gx}, \ddot{u}_{Gy}, \ddot{u}_{Gz})$ consists of two parts: (i) acceleration of the moving origin A, denoted as $\ddot{u}_A = (\ddot{u}_{Ax}, \ddot{u}_{Ay}, \ddot{u}_{Az})$, which represents kinematic excitation of the absorber by the movement of the structure, and (ii) acceleration of the ball centre G with respect to the point A being given by $\rho \ddot{u}_C$. Components of acceleration $\ddot{u}_C$ can be deduced when relations Eq. (3) are differentiated:

$$\ddot{u}_{Gx} = \ddot{u}_{Ax} + \rho \frac{d}{dt}(\omega_y(u_{Cz} - R) - \omega_z u_{Cy}),$$

$$\ddot{u}_{Gy} = \ddot{u}_{Ay} + \rho \frac{d}{dt}(\omega_z u_{Cx} - \omega_x(u_{Cz} - R)), \qquad (4)$$

$$\ddot{u}_{Gz} = \ddot{u}_{Az} + \rho \frac{d}{dt}(\omega_x u_{Cy} - \omega_y u_{Cx}), \qquad \ddot{u}_{Az} = 0.$$

After substituting Eq. (4) into Eq. (1), the Appell–Gibbs function gets a form $S = S_2 + S_1 + S_0$, where S_2, S_1 and S_0 are polynomials of the second, first and zero degree of w and $\ddot{u}_A$ components. The *reduced Appell–Gibbs function* is defined as $S^r = S_2 + S_1$. The term S_0 can be omitted because it disappears during differentiation with respect to w or $\ddot{u}_A$ components.

The function S' enables to formally write the Appell–Gibbs differential system:

$$\partial S'/\partial \dot{\omega}_x = F_{Gx}, \quad \partial S'/\partial \ddot{u}_{Ax} = F_{Ax},$$
$$\partial S'/\partial \dot{\omega}_y = F_{Gy}, \quad \partial S'/\partial \ddot{u}_{Ay} = F_{Ay}, \tag{5}$$
$$\partial S'/\partial \dot{\omega}_z = F_{Gz},$$

where $\mathbf{F}_G = (F_{Gx}, F_{Gy}, F_{Gz})$ and $\mathbf{F}_h = (F_{Ax}, F_{Ay}, F_{Az})$ are the external forces or moments acting in points G and A, respectively.

2.3 External forces

The right sides of Eq. (5) can be determined using the virtual displacements principle. In the discussed case, they include: (i) gravity forces acting in point G, (ii) external excitation in point A, (iii) influence of the lower part of the structure below point A, and (iv) dissipation forces in contact point C.

i. *Gravity forces*: Forces $\mathbf{F}_{Gg}$ originate from the vector of gravity $(0, 0, -mg)$. The elementary work performed by force $\mathbf{F}_{Gg} = (F_{Ggx}, F_{Ggy}, F_{Ggz})$ along displacement $\delta \mathbf{u}_G (\delta u_{Gx}, \delta u_{Gy}, \delta u_{Gz})$ can be expressed as

$$\delta W_{Gg} = 0 \cdot \delta u_{Gx} + 0 \cdot \delta u_{Gy} - mg \cdot \delta u_{Gz} \tag{6}$$

Virtual displacement δu_{Gz} can be determined using the third non-holonomic constraint in Eq. (3). Denoting by $\delta \varphi_-$ the virtual increments of individual components ω_-, it holds that

$$\delta u_{Gz} = \rho \left(u_{Cy} \delta \varphi_x - u_{Cx} \delta \varphi_y \right) \tag{7}$$

and therefore

$$\delta W_{Gg} = -mg\rho \left(u_{Cy} \delta \varphi_x - u_{Cx} \delta \varphi_y \right). \tag{8}$$

At the same time, the elementary work can be expressed in terms of virtual increments:

$$\delta W_{Gg} = F_{Ggx} \delta \varphi_x + F_{Ggy} \delta \varphi_y + F_{Ggz} \delta \varphi_z. \tag{9}$$

Comparison of coefficients at respective virtual components $\delta \varphi_-$ for x, y, z gives

$$F_{Ggx} = -\rho mg \cdot u_{Cy}, \quad F_{Ggy} = \rho mg \cdot u_{Cx}, \quad F_{Ggz} = 0. \tag{10}$$

ii. *External excitation in the point A*: Excitation force $\mathbf{\Phi}_A$ is considered in the horizontal direction. In the meaning of the virtual work, it acts along the displacement: $\delta \mathbf{u}_A = (\delta u_{Ax}, \delta u_{Ay}, \delta u_{Az})$. Elementary works performed by excitation forces acting in point A can be written as follows:

$$\delta W_{Ah} = \Phi_{Ax} \cdot \delta u_{Ax} + \Phi_{Ay} \cdot \delta u_{Ay} + 0 \cdot \delta u_{Az}, \tag{11}$$

where Φ_{Ax}, Φ_{Ay} are components of the horizontal excitation force. When comparing the relevant components of the elementary works the following relation arises:

$$F_{Ahx} = \Phi_{Ax}, \quad F_{Ahy} = \Phi_{Ay}, \quad F_{Ahz} = 0. \tag{12}$$

iii. *Influence of the lower part of the structure*: The force acting in point A consists of the stiffness and damping parts. Both are working on the identical virtual displacements δu_A as in the previous paragraph. Therefore, the relevant force components can be written as:

$$F_{Alx} = -C_x u_{Ax} - 2b_x \dot{u}_{Ax}, \quad F_{Aly} = -C_x u_{Ay} - 2b_y \dot{u}_{Ay}, \quad F_{Alz} = 0. \tag{13}$$

iv. *Dissipation forces in the contact point C*: The influence of damping in this case is rather complicated having a character between viscose force and dry friction. However, it can be modelled on a qualitative basis to prevent any non-pervious formulation of the model. With respect to real configuration of a structure, the damping effects are evidently sub-critical and, therefore, simplifications of its internal mechanism can be adopted. Supposing that no slipping arises in the contact, the dissipation process can be approximated as proportional to relevant components of the angular velocity vector w and the quality of the cavity/ball contact. The material coefficients characterising the rolling movement of the ball can be considered constant regardless of the direction in the tangential plane to the cavity in point C. The spin of the ball is related rather with a dry friction. Nevertheless, the influence of this damping force is even smaller than those acting in tangential directions and, therefore, such an approximation is acceptable.

Consequently, with reference to [26], the resistance force can be assumed proportional to components of the respective angular speeds. Thus, the damping forces in directions x, y, z can be defined as

$$\left(D_{Gx}, D_{Gy}, D_{Gz}\right)^T = \mathbf{T}_c \cdot \mathbf{A} \cdot \mathbf{T}_c^T \cdot \left(\omega_x, \omega_y, \omega_z\right)^T, \tag{14}$$

where $\mathbf{T}_c$ is the transformation matrix from the local coordinate system of the ball to the moving coordinates and matrix $\mathbf{A}$ reflects the damping coefficients for rolling (α) and spinning (β):

$$\mathbf{T}_c = \begin{pmatrix} \dfrac{u_{Cx}(R - u_{Cz})}{R\nu}, & \dfrac{u_{Cy}(R - u_{Cz})}{R\nu}, & \dfrac{\nu}{R} \\[2mm] \dfrac{-u_{Cy}}{\nu}, & \dfrac{u_{Cx}}{\nu}, & 0 \\[2mm] \dfrac{-u_{Cx}}{R}, & \dfrac{-u_{Cy}}{R}, & \dfrac{R - u_{Cz}}{R} \end{pmatrix}, \qquad \begin{matrix} \mathbf{A} = \text{diag}(\alpha, \alpha, \beta), \\[2mm] \nu^2 = u_{Cx}^2 + u_{Cy}^2. \end{matrix} \tag{15}$$

Finally, the external forces can be summarised as follows:

$$\begin{aligned} F_{Gx} &= -\rho m g \cdot u_{Cy} + D_{Gx} \\ F_{Gy} &= \rho m g \cdot u_{Cx} + D_{Gy} \\ F_{Gz} &= \qquad\qquad D_{Gz} \\ F_{Ax} &= F_{Ahx} + F_{Alx} = \Phi_{Ax} - C_x u_{Ax} - 2b_x \dot{u}_{Ax} \\ F_{Ay} &= F_{Ahy} + F_{Aly} = \Phi_{Ay} - C_y u_{Ay} - 2b_y \dot{u}_{Ay} \end{aligned} \tag{16}$$

2.4 Governing differential system

When the derived quantities are introduced into Eq. (5), the resulting system, together with the left part of Eq. (3), includes eight differential equations for eight unknowns $u_{Cx}, u_{Cy}, u_{Cz}, \omega_x, \omega_y, \omega_z, u_{Ax}, u_{Ay}$. Using the geometric relations

$$u_{Cx}\dot{\omega}_x + u_{Cy}\dot{\omega}_y + (u_{Cz} - R)\dot{\omega}_z = 0 \quad \text{and} \quad u_{Cx}^2 + u_{Cy}^2 + (R - u_{Cz})^2 = R^2, \quad (17)$$

which reflect the orthogonality of vectors $\mathbf{u}_C$ and $\dot{\omega}$ and geometric properties of the cavity, the final system reads:

$$J_s\dot{\omega}_x = \rho^2 \left((-u_{Cy}\omega_z - \omega_y(R - u_{Cz}))\Omega_1 - \frac{1}{\rho}\left((R - u_{Cz})\ddot{u}_{Ay} + gu_{Cy}\right) - \frac{D_{Gx}}{m} \right),$$

$$J_s\dot{\omega}_y = \rho^2 \left((u_{Cx}\omega_z + \omega_x(R - u_{Cz}))\Omega_1 + \frac{1}{\rho}\left((R - u_{Cz})\ddot{u}_{Ax} + gu_{Cx}\right) - \frac{D_{Gy}}{m} \right), \quad (18)$$

$$J_s\dot{\omega}_z = \rho^2 \left((u_{Cy}\omega_x - u_{Cx}\omega_y)\Omega_1 + \frac{1}{\rho}(u_{Cy}\ddot{u}_{Ax} - u_{Cx}\ddot{u}_{Ay}) - \frac{D_{Gz}}{m} \right),$$

$$m_s\ddot{u}_{Ax} = \Phi_{Ax} - 2b_x\dot{u}_{Ax} - C_xu_{Ax} + m\rho\frac{d}{dt}(\omega_y(R - u_{Cz}) + u_{Cy}\omega_z),$$

$$m_s\ddot{u}_{Ay} = \Phi_{Ay} - 2b_y\dot{u}_{Ay} - C_yu_{Ay} - m\rho\frac{d}{dt}(\omega_x(R - u_{Cz}) + u_{Cx}\omega_z), \quad (19)$$

where it has been denoted:

$$\Omega_1 = u_{Cx}\omega_x + u_{Cy}\omega_y - (R - u_{Cz})\omega_z, \quad J_s = J + m\rho^2R^2, \quad m_s = m\rho + M. \quad (20)$$

The damping forces enable to be simplified in the following way

$$D_{Gx} = a\omega_x + (\beta - \alpha)u_{Cx}\Omega_1/R^2,$$
$$D_{Gy} = a\omega_y + (\beta - \alpha)u_{Cy}\Omega_1/R^2, \quad (21)$$
$$D_{Gz} = a\omega_z - (\beta - \alpha)(R - u_{Cz})\Omega_1/R^2.$$

The quantities given by a solution to system Eqs. (3) and (18) describe behaviour of the structure with the absorber. Vector $\mathbf{u}_C$ depicts displacements of the contact point C of the ball and can be used to study its trajectories within the cavity. Vector $\mathbf{u}_A$ characterises horizontal movement of the point A, where the absorber is fixed to the structure. The detailed behaviour of the ball as a rotating body is given by angular velocities w. The time history of the ball rotation can be enumerated, if necessary, by means of the Euler angles.

It is worth emphasising that the system Eqs. (3) and (18) have a significantly expressed autoparametric character. Hence, the existence of semi-trivial solutions (STS) should be expected outside the resonance zone. However, it emerged that the STS can have a more general character than that defined, for example, in [3]. In other words, for values of bifurcation parameter w, which produce the STS either in the sub- or super-resonance zone, other solutions can also exist. It depends on a character of related bifurcation points, if the newly emerging solution branch reaches outside the autoparametric resonance zone, possibly involving more or all response components. These solutions, however, are generally not accessible from homogeneous initial conditions and should be looked for from relevant bifurcation points. We provide some details in the next section.

3. Autoparametric behaviour of the absorber

Behaviour of the ball absorber, when it is excited in one direction only, has a strong autoparametric character. It is characterised by a non-linear interaction between both components (longitudinal and lateral), when an enforced movement in one (longitudinal) direction destabilises the resting state of the other component. Depending on parameters of the system and excitation, the response may attain a periodic, quasi-periodic or chaotic character, which generally prevents the device from working properly. Alternatively, a similar autoparametric effect is used for the sake of a structure when an autoparametric absorber is installed. Designers often overlook the former effect because it involves non-linear relations between individual components. In the case of pendulum-type absorbers, this unwanted effect can be mitigated efficiently when a sufficiently large damping is applied [20]. However, due to small damping, ball-type absorbers are much more prone to this type of response.

The authors of this chapter thoroughly studied the effect of the autoparametric resonance of the ball-type vibration absorber. A number of distinctive solutions of the homogeneous system (no external excitation, various settings of the non-homogeneous initial conditions) were presented in [18]. Despite visually attractive shapes of certain solutions, the most important ones were used as limits separating solution groups of a certain character. Particular effects of a harmonic external excitation were studied in [17], namely different regimes of periodic or aperiodic responses and their stability, together with the effect of different values of damping. The most relevant results are summarised in this section. **Table 1** lists the numerical parameters used in figures and simulations.

The design procedure of a ball-type vibration absorber generally involves an assumption of small horizontal amplitudes of the ball [14]. Depending on the moment of inertia of the ball, the rotation inertia of the rolling sphere reduces the natural frequency of the ball-type vibration absorber as compared with the pendulum-type absorber of an equivalent length $(R - r)$. Similarly, the rolling motion of the spherical absorber reduces the efficiency of the device according to the value of the moment of inertia of the ball. For example, if a homogeneous sphere in the ball-type absorber should have the same effect as the pendulum absorber, its mass would have to be increased by a factor of $7/5$ with respect to the mass of the pendulum [14].

It appears, however, that an assumption of small horizontal vibrations can be violated easily in the resonance. Due to a limited damping there exists a significant probability that a movement of the ball within the cavity exceeds "small" values. It also appears that small damping enables various limit cycles to exist—at least for a limited time (see **Figure 2**). Such regular limit cycles are, of course, very sensitive to carefully selected initial conditions. Their existence, however even theoretical, emphasises the importance of sufficient damping in tuned mass absorbers. In conjunction with a spatial resonance movement, which can be induced by

	Absorber				Structure			
Parameter	R	r	m	α, β	M	C_x, C_y	b_x, b_y	g
	[m]	[m]	[kg]	[Ns m^{-1}]	[kg]	[N m^{-1}]	[Ns m^{-1}]	[m s^{-2}]
Value	1	3/4[a]	1	0.1	10	90	0.1	9.81

[a]*Value 1/4 is used in* **Figure 2**.

Table 1.
Model parameters used in figures and simulations.

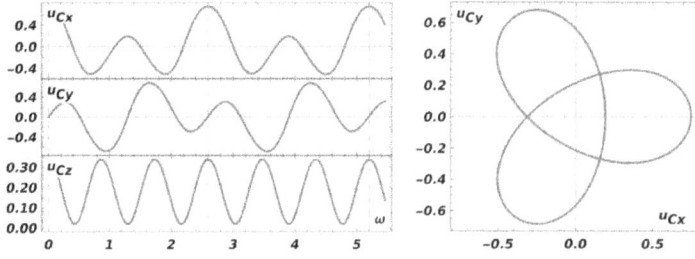

Figure 2.
Free movement of the ball for prescribed initial conditions $u_{Cx} = 0.75R, u_{Cy} = 0., \omega_x = -298.942, \omega_y = 0, \omega_z = 265.757, r = 1/4R$, no damping assumed. Left: Time history of three displacement components for two periods $T = 2.6$. Right: Trajectory of the centre of the ball in the xy plane.

uni-directional excitation, the movement of the ball in a limit cycle negatively affects the structure. It seems that the difficulty in introducing the appropriate damping is the main weakness of the ball-based absorbers and, therefore, geometrical measures should be adopted to tune the absorber.

3.1 Harmonic excitation of the cavity

In this section, we present the numerically evaluated frequency response curves for a ball-type absorber. The ball moves in a vertical plane that passes the cavity centre if an in-plane non-zero initial condition is prescribed and/or an uni-directional excitation component is applied. However, due to the non-linear character of the mathematical model, the in-plane movement is susceptible to a loss of the stability of the semi-trivial planar state for certain parameters of excitation.

Harmonic kinematic excitation (i.e., the sinusoidal form of a prescribed movement of the cavity), represents an easily understandable case, which is very convenient for both analytical and numerical treatment. It is also very popular for an assessment of dynamical properties of linear engineering structures or systems because a simple composition of the response components for individual excitation frequencies gives a realistic image of the complex response. In a non-linear case, however, this approach is not generally feasible and the frequency response curves have to be interpreted with sufficient care. Nevertheless, harmonic excitation in x direction is assumed in the following text:

$$\ddot{u}_{Ax} = u_0\omega^2 \sin(\omega t), \quad \ddot{u}_{Ay} = 0. \tag{22}$$

The initial conditions are prescribed as very small, but non-zero values in both components and the excitation amplitude are assumed as $u_0 = 0.025$. For certain excitation frequencies and amplitudes the planar response movement loses stability and lateral movement emerges. **Figure 3** shows the corresponding plots for an undamped case. The graph on the left shows the resonance curves for the longitudinal (top plot, solid curve) and lateral (bottom plot, dashed) components, obtained using a bunch of mutually independent simulation runs. The figure shows that for the selected excitation amplitude in the resonance (i) the response in the longitudinal direction increases dramatically and (ii) the zero position of the lateral component loses stability. The response in y direction attains values comparable to the longitudinal component, which represents spatial movement of the ball.

The non-linear resonance curves given by **Figure 3** have only an illustrative meaning, especially in the resonance interval $\omega \in (2.8, 3.1)$. Due to lack of damping,

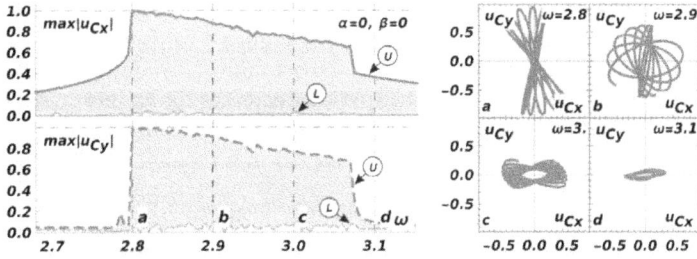

Figure 3.
*Left: The frequency response curve for longitudinal u_{Cx} (top plot, solid) and lateral u_{Cy} (bottom plot, dashed) components. Right: Projection of the contact point to the horizontal plane xy for a stabilised motion. The four plots correspond to excitation frequencies indicated by verticals in the left-hand plot. The absorber parameters are given in **Table 1**; $u_o = 0.05R$, no damping assumed, $\alpha = 0, \beta = 0$.*

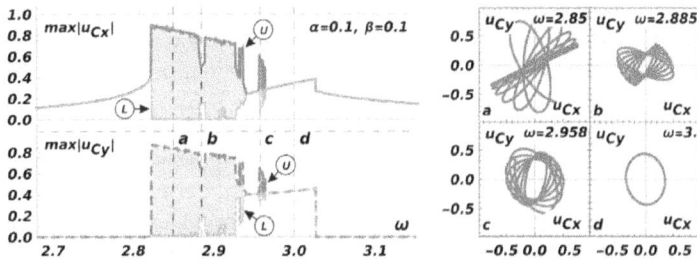

Figure 4.
*Left: Maximum (Ⓤ, blue) and minimum (Ⓛ, yellow) amplitudes of horizontal displacements. Top: Longitudinal u_{Cx} component—Solid curves; bottom: Lateral u_{Cy} component—Dashed curves. Right: Projections of the contact point of the ball to the horizontal plane xy for a stabilised motion; pictures a–d represent trajectories for frequencies indicated on the left. The absorber parameters are given in **Table 1**, $u_o = 0.05R$.*

the largest values represent only the covered integration interval. Increasing the integration time could cause a physically meaningless response.

Introduction of moderate damping between the ball and the cavity changes the resonance plots, as shown in **Figure 4**. The recorded maximal amplitudes of the stabilised response for the same excitation amplitude $u_0 = 0.025R$ are slightly lower and the resonance interval is narrower, however, the most significant change is the emergence of the stable circular or elliptic limit cycle in $\omega \in (2.94, 3.03)$.

The response remains planar outside the resonance zone in both damped and undamped cases; the lateral component vanishes. The STS is stable in the sense that for general initial conditions the trajectory stabilises in the planar state. For excitation frequencies $\omega \in (2.82, 3.03)$ the semi-trivial behaviour is unstable and the lateral movement suddenly emerges. General initial conditions for an excitation frequency in this resonance area produce a spatial response when transitional effects subside. The shape of the spatial trajectory is visible from both parts of **Figure 4**. On the left, the upper (blue, Ⓤ) and lower (yellow Ⓛ) curves in each plot correspond to maximal and minimal amplitudes of the settled response, obtained for the respective frequency ω. If both curves coincide, the response is harmonic and stationary (planar or spatial). For the non-stationary response, both curves differ, and their vertical difference indicates a width of a strip where the response takes place. If the lower curve approaches the zero value in one or both coordinates, the response at least temporarily vanishes in that coordinate (see point (a) in **Figure 4**). A positive value of the lower amplitude indicates movement in a circular strip around the vertical axis (see points (b) and (d) in **Figure 4**).

The corresponding vertical projections into the plane (u_{Cx}, u_{Cy}) of the ball trajectories are shown in the right-hand part of **Figure 4** for frequencies marked on the left by verticals (a–d). The plots (a–b) show a multi-harmonic or quasi-periodic response, where the length of the quasi-period decreases for an increasing excitation frequency. The relevant Lyapunov exponent is positive but small in this area. The very stable periodic trajectory shows plot (d) in $\omega \in (2.94, 3.03)$, with a narrow exceptional interval, plot (c).

It appears that from a certain threshold value of the excitation amplitude, the overall face of the resonance plot remains the same with the non-stationary area in the left-hand part of the resonance interval and increasing ramp representing periodic response on the right. Naturally, an increased excitation amplitude or decreased value of damping causes broadening of the resonance interval and enlargement of the response amplitude or vice versa. Changes in other parameters (m, r) influence the face of the plot more significantly, yet the overall character of the graph remains the same.

Similarly to multiple settled solutions in the resonance interval, there exist multiple solution branches also outside the resonance. They differ in stability. The best approach to their identification is an analytical way, if possible (see [15] for the case of the 2D model or [19] for a spherical pendulum). Although the unstable solutions are usually difficult to identify numerically, there are certain exceptions. **Figure 5** shows the resonance curves Ⓤ and Ⓛ from **Figure 4** enriched by two additional branches, which were obtained when the numerical simulation followed the frequency sweeping from low to high and vice versa. The sweeping process means that in every new step performed for $\omega \pm \Delta\omega$ the simulation starts from initial conditions corresponding to the final state of the previous run. This way, in fact, a small change in the driving frequency can be accommodated by the stable solution, which would be otherwise hardly accessible from random initial conditions.

The result is demonstrated in **Figure 5**. The planar response branch ①,② was obtained when the sweeping was performed from high to low, starting above the resonance interval with small but non-zero initial conditions. During continuing on the stable part of this branch above $\omega = 3.3$, curve ②, the lateral component value decreased below the machine epsilon before the resonance interval was entered and the numerical round-off then ensured continuance also on the unstable part of planar branch below $\omega = 3.3$. Here the movement remains stable with respect to perturbations the in u_{Cx} variable within the resonance interval and even further for $\omega < 2.82$, curve ①. The response is formed by a planar movement with large amplitudes. Any perturbation in the lateral direction causes a switch from ① to a generally stable planar solution ③ in $\omega \in (2.44, 2.82)$ or to a non-stationary behaviour

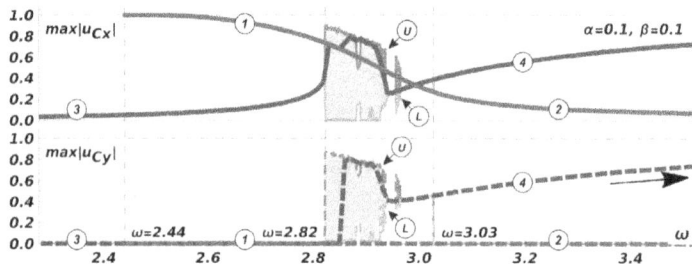

Figure 5.
Amplitudes of u_{Cx} (upper plot, solid) and u_{Cy} (bottom plot, dashed) component. Sweeping from high to low in green, ②,①; from low to high in red ③,④. The branches ①–③ exhibit the planar response; the branch ④ is spatial. The absorber parameters are as in **Figure 4.**

represented by curves ⓤ and ⓛ in the resonance interval $\omega \in (2.82, 3.03)$. Although the existence of this unstable type solution sounds theoretically, it was actually measured during the experimental examination of the spherical pendulum, see [27].

The spatial response branch ④ in **Figure 5** can be identified when starting simulation in the resonance interval, for example at $\omega = 2.85$, from small but non-zero initial conditions; sweeping the excitation frequency upwards enables the response to continue with the circular type response outside the resonance interval. The stability of the periodic trajectory gradually decreases in term of a sensitivity to perturbations, cf. [17] for details, however, the sweeping process itself is able to continue up to physically meaningless frequency values. This effect is indicated by an arrow on the right in **Figure 5**. The maximal approach of the circular trajectory to the equator of the cavity occurred for $\omega = 9.7$; for ω increased further the amplitudes start to very slowly decrease.

It is worth noting that such a periodic high-energy trajectory may represent a serious danger to the structure. Although this regime is not accessible easily, the numerical experiments show that it is unfavourably stable against perturbations in the excitation frequency and amplitude—at least for lower excitation frequencies. The spatial response in the resonance area attains also large amplitudes, however, they are not synchronised and so this case could even help to dissipate the vibrational energy to modes that are not excited by the primary loading. The planar periodic motion exhibiting high in-plane amplitudes synchronised with excitation in the sub-resonance zone may also represent a possibly dangerous state, but this effect quickly attenuates whenever the lateral component gains a non-zero value.

3.2 Random excitation of the cavity

If the harmonic excitation can be regraded as the most simple excitation case, the opposite extreme is a completely random case. For this purpose, a stationary random process is generally used, which is described by a spectral density matrix and an underlining—preferably Gaussian—probability distribution. For the sake of simplicity, only the white noise excitation will be assumed in this section. For details and more complex examples, see [28]. This simplified case of random excitation was used to assess the possibility of emergence of the high-energy spatial response due to an ambient broadband noise.

When dealing with non-linear models, the results of simulation are generally not Gaussian even for normally distributed inputs. This applies also to this case and, consequently, the results have to be represented by an estimate of a (time-dependent) probability distribution. Histograms are used for this purpose in this work.

The spatial response of the upper part of branch ④ in **Figure 5** for deterministic harmonic excitation is periodic and the relevant trajectory intersects the coordinate axes always in the same points. When random excitation is assumed, solution trajectories deviate from an ideal ellipse depending on a variance of the random process. Positions of intersections of trajectories with the coordinate axes then represent a random variable, distribution of which characterises the stochastic response. For deterministic excitation, the histograms would be concentrated in values corresponding to intersection points of the elliptic trajectory and both axes. When random perturbation of a harmonic input increases, the centre of gravity of the histogram becomes blurred. A further increase in the random perturbation intensity may cause a change of the type of the response and a switch to the lower solution branch, which is characterised by a negligible value of the lateral component and a non-zero value of the longitudinal component that reflects the relevant amplitude.

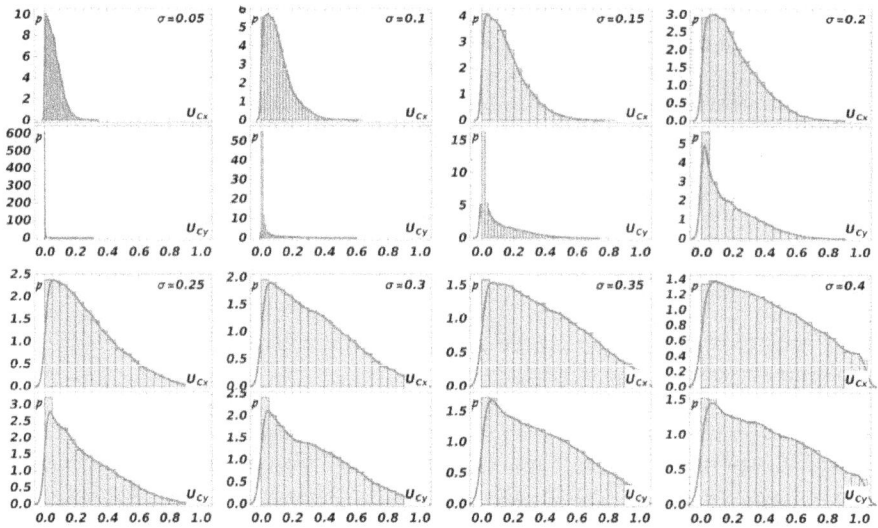

Figure 6.
*The probability density estimates for components $u_{Cx}|_{y=0}$ and $u_{Cy}|_{x=0}$ for $t \in (400, 600)$ and increasing white noise intensity σ. The absorber parameters are given in **Table 1**.*

From the simulation, it can be concluded that the spatial response may emerge depending on the variance of the input random process. This result follows from **Figure 6**, which shows probability density estimates for components $u_{Cx}|_{y=0}$ and $u_{Cy}|_{x=0}$ for an increasing white noise intensity; each simulation begins from "small" initial conditions $u_{Cx}(0) = u_{Cy}(0) = 0.01$ and counts axes crossings for the both components. In order to neglect the transient effects, the initial part of each simulation is not taken into account and only the time interval $t \in (400, 600)$ is considered. **Figure 6** shows that starting from the white noise intensity $\sigma = 0.15$, the lateral component becomes positive and for $\sigma \geq 0.35$ is the random response almost symmetric in the both components. However, the elliptic periodic response, which is typical for the spatial branch for $\omega > 2.94$, does not appear dominant in any histogram.

The random simulation was performed using the Itô version of the modified stochastic Euler method, [29], with $\Delta t = 2^{-6}$. The computation was restarted 240 times. Approximately 100 axes crossings were counted in each simulation for $t \in (400, 600)$, which number gives in total ca. 2.4×10^5 samples for each histogram.

4. Interaction of the structure and the absorber

Frequency response curves serve as a main evaluation tool when regards an efficiency estimate of the absorber. It was already shown using the analytical tools —which are available for the 2D simplified case—that the shape of such non-linear frequency response curves may be fairly complicated, see [15]. The illustrative simulation results regarding the complete equation system Eqs. (3) and (18) will be presented in this section.

The non-linear frequency response curves are shown in **Figure 7** for multiple settings of the absorber and excitation frequency. The reference data of the sample structure and the absorber used during simulation are given in **Table 1**, which setting correspond the natural frequency of the structure $\omega_0 = 3$ and an appropriate choice of the ball size $r = 3/4$. For simplicity, the damping coefficients are set equal for the absorber, $\alpha = 0.1, \beta = 0.1$, and also for the structure, $b_x = 0.1, b_y = 0.1$. The harmonic forcing is supposed in the form

$$\Phi_{Ax} = F_0 \sin \omega t, \qquad \Phi_{Ay} = 0, \tag{23}$$

where the forcing amplitude F_0 varies between 0.1 and 0.7.

The resonance curve of the linear model of the supporting structure without an absorber is shown in each plot in **Figure 7** as the black dashed curve. For cases with the absorber, the blue solid line indicates the amplitude of the structure response in

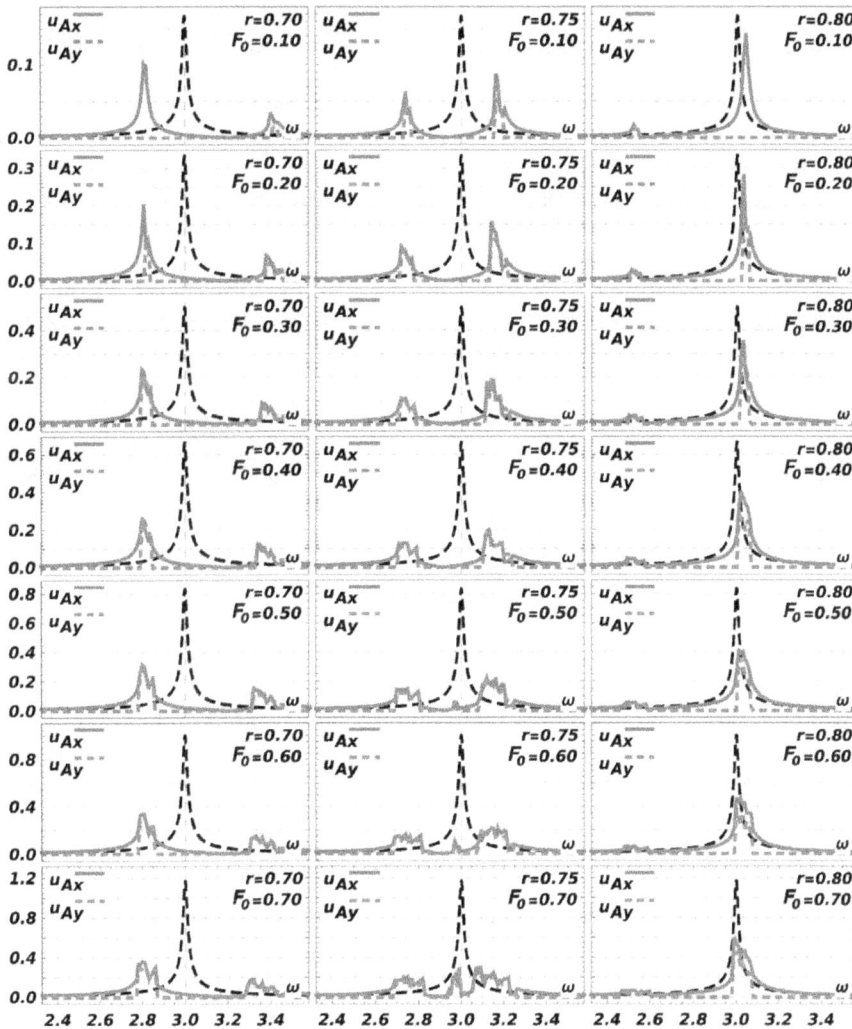

Figure 7.
*Frequency response curves of the structure equipped with the ball-type absorber. In columns—Left: $r = 0.7R$; middle: $r = 0.75R$—The optimal value; right: $r = 0.8R$. In rows: From top to bottom, the excitation amplitude $F_0 = 0.1, \dots, 0.7$. Black dashed: Frequency response of the linear structure without an absorber, blue solid and red dotted curves denote frequency response of the structure with the absorber in longitudinal and lateral components, respectively. Model parameter of the absorber and structure are given in **Table 1**.*

the longitudinal direction, u_{Ax}, the red dotted curve corresponds to the lateral direction, u_{Ay}. Three columns show the response properties of three radii of the ball: $r = 0.7R, 0.75R, 0.8R$ for the left, the middle and the right column, respectively. Finally, each row shows the response for a particular value of the excitation amplitude: $F_0 = 0.1, \dots, 0.7$.

The plots in **Figure 7** show that in the depicted case the non-zero amplitude of the lateral component arises even for the lowest forcing amplitude, namely for the case of a maximal efficiency of the absorber ($r = 0.75R$). This effect is naturally dependent on the physical properties of the structure, namely on its rigidity. In most cases are the maximal amplitudes of the both components comparable and the originally unidirectional vibration transforms into a spatial movement of the structure. See [18] for details. As the excitation amplitude increases, an additional peak emerges in the resonance frequency of the structure besides the both side extremes. This peak is significantly lower than that originating in the linear resonance, however, it appears in the both directions. Comparison of all three columns illustrates the fact that the efficiency as a function of the tuning of the absorber in terms of the radius of the ball deteriorates for $r > 0.75R$ faster than for $r < 0.75R$. Although this effect becomes less noticeable when the ratio r/R is getting smaller, in real cases it would be safer to underestimate the radius of the ball than the opposite.

Character of the responses of both the ball and the structure in the autoparametric-resonance regions is mostly quasi-periodic or chaotic. Some basic properties are evident from **Figure 8**. For a single forcing amplitude $F_0 = 0.5$ are shown the frequency response curves of components $u_{Cx}, u_{Cy}, u_{Ax}, u_{Ay}$ (four rows) in three columns for three radii of the ball: $r = 0.70R, 0.75R, 0.80R$. There are two curves in each plot which indicate (non-)stationarity of the response; the upper (blue) shows maximal amplitudes for a given forcing frequency, the lower (yellow) corresponds to minimal ones, cf. description of **Figure 4**. The plots are grouped to vertically stacked pairs. The response of the structure is shown in the second row, i.e., variables u_{Ax}, u_{Ay} and, for the sake of comparison, the response of the ball, u_{Cx}, u_{Cy}, is in

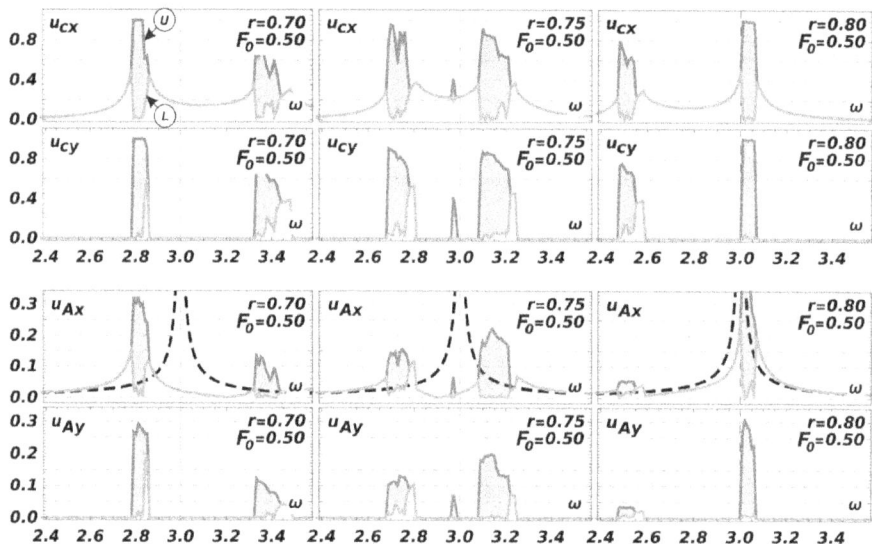

Figure 8.
Detailed frequency analysis of the response of the ball (top row) and the structure (bottom row) for three radii $r = 0.7R, 0.75R, 0.8R$ in three columns. Plots for lateral and longitudinal components are vertically stacked. Model parameters as in Figure 7.

the first row. Part of the relevant linear resonance curve is shown in the row for u_{Ax} as the dashed black curve.

It can be seen that the spatial response is mostly non-stationary. The most noticeable exception is a hardly visible interval $\omega \in (2.78, 2.80)$ for $r = 0.75R$, where the minimal and maximal response curves are non-zero and coincide for all four variables; it means that the ball and the structure move in elliptic curves. It is, however, interesting that whereas in the ball movement is dominant the lateral direction $(u_{Cy} > u_{Cx})$, for the structure is the dominant component the longitudinal one $(u_{Ay} < u_{Ax})$. Another example of such a behaviour is for $\omega \in (3.22, 3.24)$. There is one such interval for $r = 0.70R$ in frequencies above resonance $\omega \in (3.44, 3.48)$ and for $r = 0.80R$ in frequencies below resonance: $\omega \in (2.55, 2.58)$.

It is also worth noting that the movement of the ball for the depicted case $F_0 = 0.5$ reach the equator of the cavity when the radius of the ball is not optimal $(r = 0.7, 0.8)$. This case should be considered as unacceptable in a real device. However, it appears that even in this case the absorber is able to work for the sake of the structure.

The colour map plots in **Figure 9** show the sensitivity of the maximal response of the structure on the radius of the ball (vertical axis) and the loading frequency (horizontal axis). The coloured spots in both plots correspond to positions of extremes of frequency response curves in **Figure 8** for different values of the radius r. The value $r = 0.75R$, which corresponds to cases shown in the middle column of **Figure 8**, is indicated by the horizontal dashed line. Two observations are worth mentioning. The first regards position of one or both extremes when the tuning of the absorber is changing (variable r). Whereas the upper (right) extreme of the longitudinal variable decreases in magnitude and moves to higher frequencies for r decreasing from 0.75R, the position of the lower one remains stable and its value increases. For r increasing from 0.75R, the lower (left) extreme vanishes and the position of the upper one increasingly coincides with the resonance frequency of the structure. This behaviour is natural because for $r \to R$ the absorber ceases to work. The amplitude of the structure is maximal. Similar behaviour is visible also for the lateral component in the right-hand plot.

The other observation supports the previously mentioned remark regarding sensitivity of the absorber efficiency to the radius of the ball. The gradient of the response amplitudes is significantly steeper when moving up from the level $r = 0.75R$.

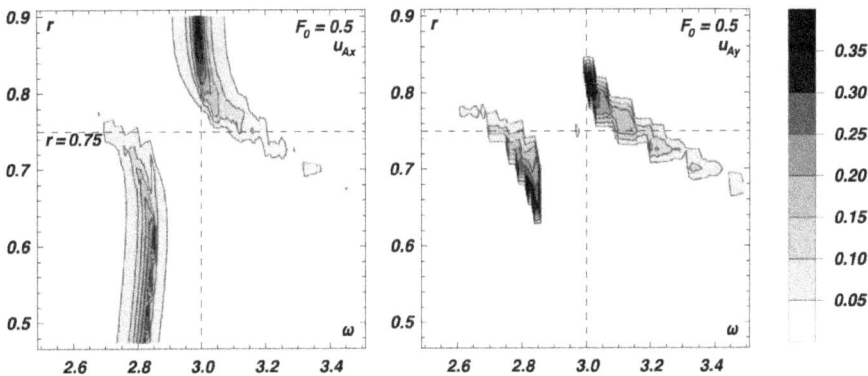

Figure 9.
Dependence of the maximal response of the structure on the radius of the ball r and the loading frequency. Left: Longitudinal component u_{Ax}. Right: Lateral component u_{Ay}. Model parameters as in Figure 7, $F_0 = 0.5$.

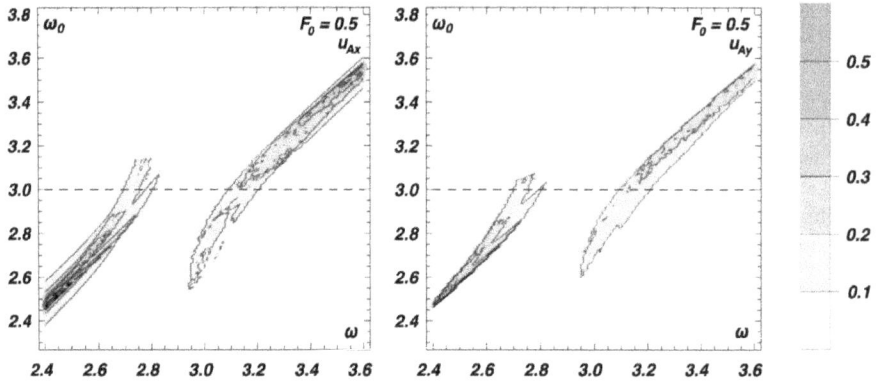

Figure 10.
Dependence of the maximal response of the structure on the natural frequency of the structure and the loading frequency. Left: Longitudinal component u_{Ax}. Right: Lateral component u_{Ay}. Model parameters as in **Figure 7**, $F_0 = 0.5$.

Similar information is provided by **Figure 10**. The ball radius (i.e., the natural frequency of the absorber) is fixed in this case to $r = 0.75R$ and the natural frequency of the structure is changing in the interval $\omega_0 \in (2.3, 3.8)$. The frequency $\omega_0 = 3$ used in **Figures 7–9** is indicated by the horizontal dashed line. It passes both extreme areas in places where the amplitudes are relatively small, a situation that corresponds to the setting shown in the middle column of plots in **Figure 7**, row for $F_0 = 0.5$.

5. Conclusions

The tuned mass absorbers are supposed to work in semi-trivial mode, avoiding any type of the autoparametric resonance effects described in this chapter. They are traditionally designed using a simplified linear, or non-linear but planar, approach, which is adequate to such an expected behaviour. However, the lack of sufficient damping makes the ball-type vibration absorbers prone to unwanted autoparametric effects, which stem from the non-linear character of the system. Thus correct and safe design has to consider possible occurrence of the autoparametric resonance. To facilitate this procedure, the non-linear mathematical model of the ball-type absorber was presented and analysed in connection to a linear model of an elastic supporting structure. The model of the absorber consists of six degrees of freedom constrained by three non-holonomic relations. The complete system with the structure comprises ten first-order ordinary differential equations.

It was shown that in systems with small damping, the desired planar STS is prone to loss of stability even for small excitation amplitudes. This danger increases with increasing excitation amplitude. Although the resonance interval is relatively narrow, the spatial response of the absorber can emerge also due to a broadband random excitation, provided that the intensity of the random noise exceeds a certain limit. The spatial movement of the ball within the absorber is unfavourably stable with respect to random perturbations that correlate with the resonance frequency of the structure.

The efficiency of the absorber is obviously dependent on a proper tuning. It was shown that the absorber efficiency deteriorates faster if the ratio between radii of the ball and the cavity is greater than the optimal one, rather than in the opposite case. Although this effect becomes less noticeable when the ratio r/R is getting smaller, in real cases it would be safer to underestimate the radius of the ball with respect to the cavity, rather than the opposite.

Although the described resonance state should be preferably avoided, it appears, however, that a limited induced lateral movement of the ball may help to dissipate the harmonic loading energy and stabilise the structure. However, this mechanism should not be relied upon in the design procedure as it can set off movement in dangerous non-linear high-energy limit cycles.

Both cases of harmonic and random excitation indicate a need for further investigation of the topic. A more thorough parametric study should comprise different system parameters and structure types in the case of harmonic loading. A deeper stochastic analysis is also necessary, which should comprise the effect of a supporting structure. Nevertheless, it can be concluded that autoparametric resonance effects may be encountered in practice more often than expected. This can be dangerous for structures if adequate countermeasures are not applied.

Acknowledgements

The kind support of Czech Scientific Foundation No. 19-21817S and RVO 68378297 institutional support are gratefully acknowledged.

Author details

Jiří Náprstek[*†] and Cyril Fischer[†]
Institute of Theoretical and Applied Mechanics of the Czech Academy of Sciences, Prague, Czech Republic

*Address all correspondence to: naprstek@itam.cas.cz

† These authors contributed equally.

IntechOpen

References

[1] Elias S, Matsagar V. Research developments in vibration control of structures using passive tuned mass dampers. Annual Reviews in Control. 2017;**44**:129-156

[2] Sun JQ, Jolly MR, Norris MA. Passive, adaptive and active tuned vibration absorbers—A survey. Journal of Mechanical Design. 1995;**117**(B): 234-242

[3] Tondl A, Ruijgrok T, Verhulst F, Nabergoj R. Autoparametric Resonance in Mechanical Systems. Cambridge: Cambridge University Press; 2000

[4] Haxton RS, Barr ADS. The autoparametric vibration absorber. Journal of Engineering for Industry. 1972;**94**(1):119-125

[5] Nabergoj R, Tondl A, Virag Z. Autoparametric resonance in an externally excited system. Chaos, Solitons & Fractals. 1994;**4**(2):263-273

[6] Miles JW. Stability of forced oscillations of a spherical pendulum. Quarterly Journal of Applied Mathematics. 1962;**20**(1):21-32

[7] Chen J, Georgakis CT. Tuned rolling-ball dampers for vibration control in wind turbines. Journal of Sound and Vibration. 2013;**332**(21):5271-5282

[8] Cui W, Ma T, Caracoglia L. Time-cost "trade-off" analysis for wind-induced inhabitability of tall buildings equipped with tuned mass dampers. Journal of Wind Engineering and Industrial Aerodynamics. 2020;**207**: 104394

[9] Routh E. Dynamics of a System of Rigid Bodies. New York: Dover Publications; 1905

[10] Bloch AM, Marsden JE, Zenkov DV. Nonholonomic dynamics. Notices of the American Mathematical Society. 2005; **52**:324-333

[11] Hedrih K. Rolling heavy ball over the sphere in real Rn3 space. Nonlinear Dynamics. 2019;**97**(1):63-82

[12] Udwadia FE, Kalaba RE. The explicit Gibbs-Appell equation and generalized inverse forms. Quarterly of Applied Mathematics. 1998;**56**(2): 277-288

[13] Borisov AV, Mamaev IS, Kilin AA. Rolling of a ball on a surface. New integrals and hierarchy of dynamics. Regular and Chaotic Dynamics. 2002; **7**(2):201-219

[14] Pirner M, Fischer O. The development of a ball vibration absorber for the use on towers. Journal of the International Association for Shell and Spatial Structures. 2000;**41**(2):91-99

[15] Náprstek J, Fischer C, Pirner M, Fischer O. Non-linear model of a ball vibration absorber. In: Papadrakakis M, Fragiadakis M, Plevris V, editors. Computational Methods in Applied Sciences. Vol. 2. Dordrecht: Springer Netherlands; 2013. pp. 381-396

[16] Legeza VP. Determination of the amplitude-frequency characteristic of the new roller damper for forced oscillations. Journal of Automation and Information Sciences. 2002;**34**(5–8): 32-39

[17] Náprstek J, Fischer C. Stable and unstable solutions in autoparametric resonance zone of a non-holonomic system. Nonlinear Dynamics. 2020; **99**(1):299-312

[18] Náprstek J, Fischer C. Limit trajectories in a non-holonomic system of a ball moving inside a spherical cavity. Journal of Vibration Engineering & Technologies. 2020;**8**(2):269-284

[19] Náprstek J, Fischer C. Auto-parametric semi-trivial and post-critical response of a spherical pendulum damper. Computers and Structures. 2009;**87**(19–20):1204-1215

[20] Pospíšil S, Fischer C, Náprstek J. Experimental analysis of the influence of damping on the resonance behavior of a spherical pendulum. Nonlinear Dynamics. 2014;**78**(1):371-390

[21] Borisov AV, Fedorov YN, Mamaev IS. Chaplygin ball over a fixed sphere: An explicit integration. Regular and Chaotic Dynamics. 2008;**13**(6): 557-571

[22] Matta E, De Stefano A, Spencer BF Jr. A new passive rolling-pendulum vibration absorber using a non-axial guide to achieve bidirectional tuning. Earthquake Engineering and Structural Dynamics. 2009;**38**:1729-1750

[23] Legeza VP. Numerical analysis of the motion of a ball in an ellipsoidal cavity with a moving upper bearing. Soviet Applied Mechanics. 1987;**23**(2): 191-195

[24] Matta E. A novel bidirectional pendulum tuned mass damper using variable homogeneous friction to achieve amplitude-independent control. Earthquake Engineering and Structural Dynamics. 2019;**48**(6):653-677

[25] Awrejcewicz J, Cheaib A, Losyeva N, Puzyrov V. Responses of a two degrees-of-freedom system with uncertain parameters in the vicinity of resonance 1:1. Nonlinear Dynamics. 2020;**101**(1):85-106

[26] Náprstek J, Fischer C. Appell-Gibbs approach in dynamics of non-holonomic systems. In: Reyhanoglu M, editor. Nonlinear Systems. Rijeka: IntechOpen; 2018. pp. 3-30. Available from: https://doi.org/10.5772/intechopen.76258

[27] Pospíšil S, Fischer C, Náprstek J. Experimental and theoretical analysis of auto-parametric stability of pendulum with viscous dampers. Acta Technica. 2011;**56**(4):359-378

[28] Fischer C, Náprstek J. Numerical solution of a stochastic model of a ball-type vibration absorber. In: Chleboun J, Kůs P, Přikryl P, Rozložník M, Segeth K, Šístek J, et al., editors. Programs and Algorithms of Numerical Mathematics 20. Prague: Institute of Mathematics CAS; 2021. pp. 40-49.

[29] Kloeden PE, Platen E. Numerical Solution of Stochastic Differential Equations. Berlin-Heidelberg: Springer; 1992

www.ingramcontent.com/pod-product-compliance
Lightning Source LLC
Chambersburg PA
CBHW081231190326
41458CB00016B/5747